This book provides a discussion of animal vocal communication that avoids human-centered concepts and approaches, and instead links communication to fundamental biological processes. It offers a new conceptual framework – assessment/management – that allows for the integration of detailed proximate studies of communication with an understanding of evolutionary perspectives. This framework is distinguished by two central features: (1) it places self-interested assessment on a par with the production (management) side of communication; and (2) it highlights regulation of the behavior of other individuals as the key process in management. Animals use signals in self-interested efforts to manage the behavior of others, and they do so by exploiting the active assessment processes of other individuals. The authors contend that it is the interplay between assessment and management that underlies the production and functioning of animal communication systems. Communication reflects the fundamental processes of regulating and assessing the behavior of others, not of exchanging information. This will be a landmark text for all those interested in animal communication.

ANIMAL VOCAL COMMUNICATION: A NEW APPROACH

ANIMAL VOCAL COMMUNICATION: A NEW APPROACH

DONALD H. OWINGS

Department of Psychology
University of California
Davis, California 95616-8686
USA

and

EUGENE S. MORTON

National Zoological Park
Smithsonian Institution
Front Royal, Virginia 22630
USA

CAMBRIDGE
UNIVERSITY PRESS

PUBLISHED BY THE PRESS SYNDICATE OF THE UNIVERSITY OF CAMBRIDGE
The Pitt Building, Trumpington Street, Cambridge CB2 1RP, United Kingdom

CAMBRIDGE UNIVERSITY PRESS
The Edinburgh Building, Cambridge CB2 2RU, United Kingdom
40 West 20th Street, New York, NY 10011-4211, USA
10 Stamford Road, Oakleigh, Melbourne 3166, Australia

First published 1998

Printed in the United Kingdom at the University Press, Cambridge

Typeset in 10/13 Times [KWS]

A catalogue record for this book is available from the British Library

Library of Congress Cataloguing in Publication data
Owings, Donald H. (Donald Henry), 1943–
   Animal vocal communication: a new approach/Donald H. Owings,
Eugene S. Morton.
       p.   cm.
   Includes bibliographical references (p.   ) and index.
   ISBN 0 521 32468 8 (hardback)
   1. Animal communication. I. Morton, Eugene S. II. Title.
QL776.085   1998
591.59′4–dc21   97-27049 CIP

ISBN 0 521 32468 8 hardback

Contents

Preface

Animal Vocal Communication: A New Approach results from our belief that the field of animal communication needs new blood. We have tried, independently, to convince colleagues that the informational perspective is not adequate as a concept or methodology to understand either the evolution or the process of vocal communication. In 1987, we began to collaborate on this book, deciding on the order of authors with the toss of a coin.

We have attempted to include in this book those aspects of theory that we feel have 'staying power,' i.e., those that suggest and provide the best questions and will not lead to blind alleys. Game theory, in particular, has increased in popularity. It deals with frequency dependence – the concept that the success of an animal's behavior depends on the relative frequency with which it occurs in the animal's own population. A game-theoretic approach can be contrasted with studies of adaptation that deal with structure, function and with nonsocial activities as well as with social behavior. Adaptation suggests optimization, such as the optimal shape of a bird's wing for soaring flight. But both of these approaches have staying power; they are complementary to each other rather than in conflict. A behavioral strategy will not work if someone else can do it better because they have the better structural design. It is the latter that affects the outcome of evolutionary competition. We feel that a better approach to vocal communication is to study function, the relation between form or structure and performance. Furthermore, the social consequences of communicating can, and should be, studied empirically. We hope to provide students with the material necessary to form and test hypotheses about communication, its effects, and its evolution.

It is somewhat unexpected that two people with such different backgrounds and interests should unite to write a book on communication, and amazing that they completed it! Don Owings brings the views of a psychologist with an organismic orientation into the book. Gene Morton brings the views of a biologist who studies natural selection. We also use mammals and birds, respectively, as our main study animals. We hope that the diversity due to our backgrounds is reflected in this book.

Acknowledgments

Don Owings is indebted to the University of California at Davis for the sabbatical leave that facilitated his progress on this book. His contributions were strongly influenced by many years of interacting with students and colleagues. Most important is David Hennessy, with whom a *management* approach to animal communication was first developed. Dick Coss, Dan Leger, Matt Rowe, and Ron Swaisgood were rich sources of intellectual stimulation during many hundreds of hours of collaboration. Ragon Owings played a major role in assembling the bibliography and preparing many of the figures. Dan Leger and Pat Arrowood read the manuscript and provided many useful suggestions. Gene Morton provided fine and memorable hospitality for our writing sessions on the Severn River.

Gene Morton was provided with valuable help in compiling references and determining underlying concepts by Kim Young in the early stages of the project; Fig. 1.1 results almost entirely from her work. Kim Young's support came from The Friends of the National Zoo (FONZ). Over the years, FONZ also supported many aspects of Gene Morton's research that are incorporated in this book. When Kim moved away, Robin Yung and then Alexandra Sangmeister helped with the book. The Smithsonian Institution has generously supported Gene Morton's research for this book through grants from the Smithsonian Women's Committee, the Scholarly Studies Program, and the Research Opportunities Fund. The main support that allowed him to complete his portion of the book was a grant from the Regents Publications Program in 1993, which supported a one-year sabbatical to devote time to this project and, especially, to meet with Don Owings in California, where Don's family helped make the work enjoyable. Gene Morton's wife, Bridget Stutchbury, provided support throughout the years, and her keen scientific and editorial talents resulted in the best parts of the book attributable to him. Jane Smith corrected the many problems of detail in the text, and smoothed and clarified many convoluted passages during the process of copy editing.

Prologue

Carolina wren

A male Carolina wren was widowed in the flash of a sharp-shinned hawk's attack. It was late September, but singing and pairbonding are possible throughout the year for this wren, as is the case with tropical-living birds, even though the hawk struck in the State of Maryland, USA. The widowed male continued to sing and otherwise defend his permanent year-round territory using a repertoire of 42 song types. He sang one type 5 to 120 times before switching to another. During the two months after he became independent from his parents, his song-learning phase, he had learned 85% of his song types from neighboring males; he learned the remaining 15% while dispersing or from males more distant from his territory and not found in his neighbors' repertoires. At 9:00 a.m. a stranger wren was detected. The male approached it with plumage fluffed, quickly sang three song types without the usual pause between them, then attacked. The intruder gave high-pitched *pi-zeet* calls, an appeasement or friendly call between mates or siblings. It then gave high *pee pee pee* calls as the resident male continued to attack, producing short growls during each attack flight. Between attacks, the resident male also uttered harsh, low *chirrs* between attacks. The intruder fled silently. At 10:30 a.m. a second intruder was detected. This time the intruder responded to the resident's attack by fluffing and simultaneously calling a chattering *thirrrrrrrrrr*. Next, the new bird gave the chattering call during the resident's song rather than after it. Now the male fluffed and hopped stiffly around the intruder like a miniature strutting turkey, uttering high-pitched *tsuck* calls. The intruding female uttered *pi-zeeets* when the male approached, and chattered when he sang or might attack her. A pair bond had formed seven months before they nested in the spring.

1

Fig. P.1. A Carolina wren. Males and females have identical plumage.

The pair foraged within earshot of one another throughout the daytime hours. Only the male sang and only the female produced *dit-dits* and chatters, whereas both *pi-zeeted, chirred, rasped,* growled when attacking intruding wrens, and *chirted. Chirts* were variable in timing and in their component frequencies and were used when a hawk perched near or when a pair member was moving a long distance. Mostly the female used *chirt* sequences when a hawk remained nearby; the male foraged for food while the female kept up the surveillance. If the hawk moved, even slightly, the female's *chirts* shifted to a higher pitch than before the movement. Then, the alerted male stopped foraging. *Chirts* were combined with *pi-zeets* when one of the mates approached the other from a long distance, and they were combined with *rasps* or *chirrs* when chasing an intruder. Snakes elicited *rasps*, whereas most predators elicited *chirt* sequences.

After two more years had passed, the male was killed by a feral house-cat. Within two hours, wren pairs on either side had moved onto the territory and had chased the female away.

Anna and her mother

Anna, an energetic eleven-month-old infant, had been placed in her new, spring-suspended 'bouncy chair' for the first time, and was just mastering the skill of bouncing in it. Her mother, who was speaking briskly in her steady, low-pitched voice to a visiting neighbor, smiled as the bouncing began, slowed the pace of her words, and switched to a high-pitched, lilting tone of voice, 'WELL! Aren't YOU the smart one? HOW'D you learn to do that so FAST?' Anna paused, locking eyes with her mother, and both broke into delighted smiles as Anna kicked herself into even higher bounces, and her mother encouraged her, 'THAT'S my SMART girl; she's no slug like her MAMA was!' When her mother resumed speaking to her neighbor, switching immediately back to her 'adult' voice, Anna quickly tired of bouncing, and struggled to stand in her chair. These efforts had yielded success by the time her mother noticed; once again mother radically changed her manner of speaking, emitting a single, sharp, low-pitched 'NO!!' But, mother's prohibition was too late: Anna toppled from her chair, producing a loud thump with her head as she landed on the (mercifully carpeted) floor. Frightened and a bit hurt, Anna began to cry as her mother rushed to pick her up. As she sat in a rocking chair to begin to comfort Anna, mother adopted yet another distinctive pattern of speaking, gently shushing her crying baby, and speaking quietly, with long, smooth, low and falling pitch contours, 'Ssshhhh; thaaat's okaay; maaama's heere.' As Anna became quiet, her mother continued to rock and cuddle her and resumed her conversation with the neighbor, once again making the striking transition to the choppy, rapid-fire, monotonic pattern of speaking used when addressing other adults.

Fig. P.2. Anna and her mother.

The hooded warbler

A male hooded warbler occupied his nonbreeding territory in the Yucatan Peninsula, Mexico, from 20 September until 15 April. He used metallic-sounding *chink* calls to defend this territory, calling about once each two seconds, but uttered harsh growling sounds before chasing the (rare) intruder who did not leave his territory. Calling and foraging for food took place near and on the ground. On 15 April, he began his return trip to his breeding territory in northwestern Pennsylvania, arriving there on 10 May. Here, he not only *chipped*, but sang one song over and over from positions high in the trees of his forested territory – a loud, clear, ringing, tonal, *weeta weeta weeTEEoo*, sung at a regular cadence of 5–8/ min, faster if another male sang nearby. Other males had returned as well but no females. Some he distinguished as returning oldtimers such as himself, whereas some were new. These newcomers he responded to vigorously. If they sang near his boundary, he flew to it and displayed his black throat and drooped his wings. Rarely was fighting needed to beat newcomers.

A female entered his territory on 18 May, three days after her arrival on the breeding grounds, and began nest building on 19 May. She uttered only *chink* calls, and stayed low while foraging, just as she had on her winter territory, in Belize. As soon as she began nest building, the male stopped repeating the same song and mixed five other songs into his repertoire, singing at a faster rate, 10–12/min, especially when other males were nearby, often matching their song types during countersinging matches. He had changed from repeating a single song type, unique to him, to a mixed repertoire shared with neighboring males. Some songs used in 'mixed' singing were only found in this local population, indicating that the males had learned them after arriving here. Few of the males, and none of the females, had been raised here.

During the time she was nest building, laying eggs, and incubating them, the female called *chink* repeatedly whenever she was away from the nest. Of the four young hatched in their nest, two were fathered by a neighboring male. The male was apparently unaware of this for he fed all the nestlings at the same rate as the female fed them. A chipmunk approached the nest. The female used more rapidly uttered *chink* calls, with a higher pitch than normal. As the chipmunk climbed toward the nest, the male flew at it and the female produced very high-pitched, rapid *chinks*. Not yet capable of flight, the nestlings fluttered out of the nest, dispersing in all directions. After the chipmunk left, the parents stopped

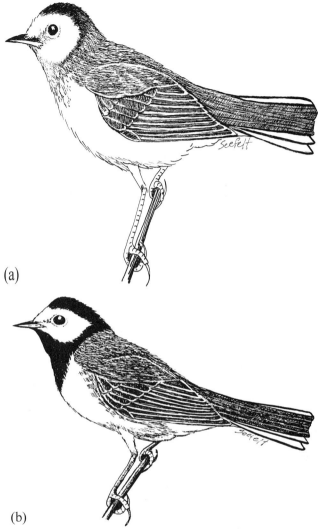

Fig. P.3. Female (a) and male (b) hooded warblers differ in plumage and only males sing; otherwise, they share the same vocalizations.

the high-pitched *chinks* and returned to foraging. The fledglings uttered high-pitched *chippity-chups*, particularly when a parent was nearby. But the male tended to feed two young and the female the other two young. After a few days, the male took on most of the feeding duties while the female began to build a second nest which would receive a new set of eggs within a week.

The Túngara frog

The male Túngara frog floated in the small pond on the floor of a clearing in a Panamanian rain forest. Although he was diminutive by absolute standards (between 3 and 5 cm long from snout to rear end), he was nevertheless one of the largest of the many males in the pond. He inflated his throat sac to a startling size, approximating that of his head and body combined, and began to produce surprisingly loud calls – *whiine; whiine; whiine*. Responding to this sound, other males in the pond began to call similarly, which in turn stimulated our male to produce a more complex vocalization – *whiine, chuck-chuck; whiine, chuck-chuck*. Soon these frogs were living up to their reputation as one of the most vocal of frog species; the moist tropical air was filled with the clamor of their *whiine, chuck-chucks*.

Fig. P.4. A fringe-lipped bat about to catch a Túngara frog.

A fringe-lipped bat was hunting on the wing nearby, and heard these vocalizations. Taking advantage of the fact that the *chuck* portions of the calls are especially easy to locate, the bat pinpointed a calling frog, swooped and caught it. Silence immediately descended upon the pond, but was short lived; a few males soon resumed calling, and the chorus once again spread throughout the pond, but this time no *chucks* were added as the chorus grew – *whiine*; *whiine*; *whiine*. A female entered the water, swimming among the calling males and stopping periodically within 10–20 cm of a male, listening to his calls. She moved on without interacting, wriggling free from one male who attempted to clasp her sexually, and finally pausing near two males who began to add *chucks* to their calls – *whiine, chuck*; *whiine, chuck*. She oriented initially toward the smaller of these two males, then toward the lower-pitched *chuck* sounds of our large male. Moving quickly to the larger male, she accepted his sexual clasp.

The female swam to the pond shore, carrying our now-silent mounted male with her, and left the pond, possibly to escape the danger from predation generated by the conspicuousness of the continuing vocal chorus. After midnight, when chorusing had ceased, the female re-entered the pond, still carrying her mate. For about an hour, she released eggs near the pond's edge and our male took each batch with his hind legs, fertilized them and created a foam nest for them by whipping their outer covering into a 'meringue'. When the nest was complete, the female swam from underneath our male, who lingered for a minute, apparently fatigued by egg-beating, before disappearing into the darkness.

The California ground squirrel and the rattlesnake

The female California ground squirrel was huddled in her underground nest with her five-week-old pups when she detected the faint sound of something moving through the dry-leaf litter near the mouth of her burrow. This was no rustling sound like a walking animal; it was the continuous *ssshh* of something sliding. Leaving her pups, she moved cautiously toward the entrance, unable to see in the burrow's darkness, but sniffing and listening intently. As she progressed, the odor of the intruder became more apparent. Rattlesnake!

The squirrel's pace slowed as she strained to pick up more precise cues. Where exactly was the snake? Was it small, or large enough to be hunting her pups, and to pose a serious threat of injury to herself? Was it cold and sluggish from the cool spring morning, or had it warmed itself to an

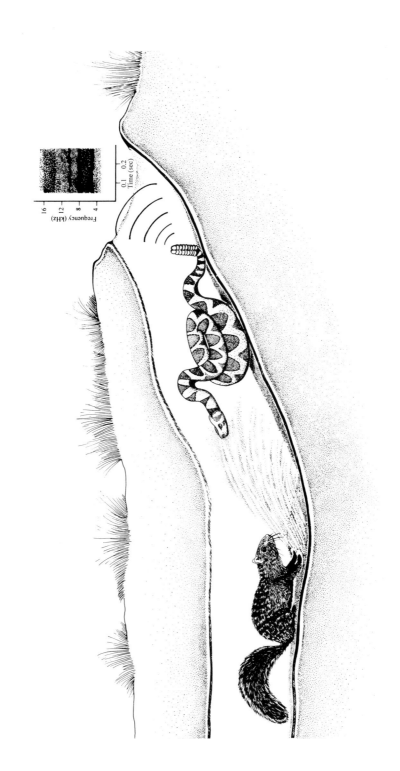

Fig. P.5. A California ground squirrel, defending her pups, assesses a rattlesnake by causing it to rattle.

action temperature in the sun? Gathering a mound of dirt in front of her, she kicked it ahead of her, paused to listen, stepped forward, and sprayed a second load of earth into the darkness. This second fusillade struck the snake full in the face, causing it to rattle as it withdrew its forequarters into a ready-to-strike defensive posture. The mother backed off quickly at the distinctively dangerous sound of a large, warm rattlesnake, stumbling over one of her pups who had followed her to the scene.

Chattering loudly, the mother returned to deal with the snake as the sound of her call sent the pup back to its nest. While she continued to vocalize repetitively, she began to construct a substantial plug of earth between herself and the snake, pausing intermittently to pack it tightly by pounding it with her head. Once the rattlesnake's access to her pups was eliminated, she stopped calling, relaxed perceptibly, and re-entered the nest. Picking up a pup in her mouth, she left her burrow via another exit and traveled to the burrow system of an adult male some 25 meters away, depositing her youngster there. As she shuttled back and forth between old and new nest sites, she showed evidence of continued concern about snakes, pausing intermittently to wave her fluffed tail from side to side in a distinctive visual signal. She did not rest until all six of her pups had been transferred to their new nest site.

The wren, the warbler, Anna's mother, the Túngara frog, and the ground squirrel often vocalized only during an important but rare event such as a fight, but at other times called almost constantly. Nevertheless, it is possible in most cases to identify discrete sounds (the individual *whine* calls of Túngara frogs, for example) that reflect vocal communication. The discrete nature of most vocalizations gives communication the appearance of a discrete event, a call and a response, rather than an ongoing process. However, it will be argued that such discrete vocalizations reflect an underlying process of regulating the vocalizer's circumstances. An analogy to feeding illustrates our point. Animals must eat, a discrete event, but this is in service of maintaining their energy balance. Effectively regulating one's energy balance and social circumstances is, in turn, a prerequisite for success in other processes, such as competing for mates and territories, and ultimately in reproducing.

Eating and vocalizing differ in an important way: eating *acquires* energy, whereas signaling by any modality – sound, sight, touch, electrical, or chemical – *costs* energy. So why is signaling so common? One answer to this question is that signaling has often been favored by natural selection because it substitutes for behavior that is even more energetically

costly, such as attacking. In addition, fighting is more likely than com-
municting to result in injury or attract predators, important costs in addition to those on the energetic balance sheet. This idea is developed further
in Chapter 3.

Signaling is similar to eating in another fundamental way: they both
reflect exploitation of a resource. Eating capitalizes upon the nutrients
available in other organisms; signaling exploits the process of assessment
by other animals. Indeed, this provides a more general reason for the
prevalence of communication; the fact that animals constantly assess
their environments (in order to make adaptive behavioral decisions)
makes it possible for individuals to use signals to manage their circumstances. A simple evolutionary example, initially formulated by Charles
Darwin, illustrates this point (see also Andrew, 1972; Krebs & Dawkins,
1984). Why do dog-like mammals intimidate rivals not only by biting
them, but also by exposing their canine teeth in a threat signal?
Retracting the lips from the teeth is a necessary preparation for biting
(so that the lips are not bitten). For this reason, retraction of the lips can
be used as a cue to assess when biting is likely to occur. The use of this
assessment cue creates a regulatory opportunity: a wolf can deal with
rivals by using the 'cheaper' threat display instead of the more costly
act of biting.

Communication is a two-way process. The behavior of each commu-
nicant is both cause and effect of the other's behavior. Each exchange is
difficult to understand in isolation. It is also difficult to consider communication in isolation from other behavior. Communication is better
understood in terms of the *regulatory* problems each participant faces:
each tries to regulate the behavior of the other. Natural selection should
favor mechanisms that are effective in minimizing the difference between
preferred and actual states of the other's behavior. An animal can regulate the behavior of another animal either through direct physical
manipulation or by communicating. Communicating with *vocalizations*
is just one means of regulating social circumstances. Concurrent changes
in posture, locomotion, and body orientation might, in a male California
ground squirrel, for example, function together to keep another male
away from an estrous female (Owings & Hennessy, 1984).

This all-inclusive view of communication differs in a fundamental way
from earlier ones. Students in a graduate seminar on animal communication were asked to give a short definition of animal communication.
Eighty-five percent defined communication as the 'transfer of information
from one animal to another.' The remaining students stated something

like 'when one animal's signal influences the behavior of another.' The more popular definition, information transfer, arises from application of a concept in wide use, the 'information concept.' The root meaning of communication is 'sharing,' suggesting that some transfer takes place, as in the information often said to be exchanged during speech. The information concept has proven very useful in making sense of animal communication (Smith, 1977), but we will argue that it has become too central, deflecting our attention from the more fundamental idea of regulation.

The concept of sharing during communication is also problematic. Smith (1977) used the term sharing in part to emphasize that information transmitted to someone else is not lost by the sender; thus, it is shared, not given away. But, the additional implication, of mutualism or co-operation, was also part of this approach, as it was for many other researchers in communication at that time (Dawkins & Krebs, 1978). The assumption that cooperation lies at the root of all communication is now recognized to be inconsistent with a basic tenet of the logic of natural selection: that conflict is just as likely as cooperation. For example, courtship and pairbonding in animals were viewed as a form of cooperation to reproduce (Trivers, 1972). Now, it is recognized that there are conflicting interests between mates and, of course, mates do not reproduce 'for the good of their species' (Williams, 1966). We now think of reproduction in terms of mixed reproductive strategies, with males and females often copulating with nonmates. Communication has changed too. The view that communication is selected to maintain orderly interactions and social structures is likewise an outdated concept. As a field of evolutionary biology, communication research must be guided by the logic of natural selection.

One way that this guidance can be made explicit is to use 'communication' as a general term for the subject, but to use 'management' to describe the part of the process that includes the production of a vocal signal. Selection favors those able to use signals to manage the behavior of others in their own interests. Other individuals perceive the signal, and respond or not to the signal in *their* best interest. Signaling and perceiving serve self-interest. The term 'perceiving' is used here, rather than the more common 'receiving,' to emphasize the active role of this process in communication. Perceiving individuals achieve their importance as a force in communication and its evolution through their 'active assessment' of signals.

The warbler, wren, mother, frog, and squirrel illustrate diverse forms of vocal communication. How do we discover general explanations for the form of the signals, the responses of perceiving individuals? Why would natural selection favor such interactions and why do they take the form they take? These questions and others are summarized in Chapter 5, after the conceptual and practical tools have been provided to understand them. Our goal is to provide both a way to think objectively about vocal communication and a way to perform research to answer the specific questions that generate your interest in communication.

1
Overview of ideas

What concepts have been used to explain the sorts of behavior described in the prologue? The purpose of this chapter is to review the variety of those ideas. The overview is historical in part, but is also organized around Nikolaas Tinbergen's four-part classification of questions that can be posed about behavior – about the evolutionary history and functions of behavior (ultimate questions), and about the immediate regulation and development of behavior (proximate questions).

This chapter does not distinguish between the ideas that do and do not form a part of a modern synthesis. Our synthesis is presented in Chapter 2, and is developed from only a portion of the concepts discussed here. The goal in Chapter 2 is to identify that subset of old and new concepts with the greatest capacity to provide an understanding of communicative behavior.

1.1 An evolutionary approach: ultimate questions
1.1.1 Natural selection and the functions of behavior

Well over a hundred years ago, Charles Darwin provided the essential elements of the evolutionary framework in which this book is cast. His principle of evolution through natural selection is very powerful, but also so elegantly simple that it can be summarized by just four points (Alcock, 1989). (1) There is variation among the individuals within a species, in body form, physiology, behavior, and so on. (2) Some of this variation is heritable; in other words, some of the distinctive characteristics of individuals can be passed on to their young, so that offspring tend to resemble their parents more than they do other members of the species. (3) Even though adults produce many offspring, populations do not consistently grow in proportion to the number of offspring produced. This

13

implies that most offspring must die without reproducing. (If most individuals did not die before reproducing, then populations of even slowly reproducing species such as humans would increase exponentially and occupy all available space on Earth; obviously, such population growth is rare, and cannot be sustained for long.) (4) Because of their special, inherited attributes, some types of individuals are likely to cope more effectively than others with problems arising from such sources as predators, competitors, and disease. These individuals will tend to leave more descendants than other members of the species with different and less successful inherited traits. Across consecutive generations, the greater reproductive success of the more successful individuals will lead to their becoming the most common types within the population.

Many of Darwin's ideas are still central in our modern understanding of behavior. A very important one is the concept of *function*, implicit in item four above. Some of the 'special inherited attributes' are tendencies to behave in particular ways; these have consequences that lead to greater reproductive success, and thus to increases in the numbers of individuals who behave that way in subsequent generations. Such consequences are now called the functions of behaving in those ways. From the perspective of natural selection, it is easy to understand, for example, how the tendency to emit calls when a predator comes on the scene might have been favored (such behavior is common during predatory encounters – see, for example, the Prologue item on the ground squirrel and rattlesnake). Such calling might help one's own offspring by inducing them to take refuge before the predator becomes a threat, thereby facilitating the offsprings' survival and subsequent reproduction. If these calls do not incidentally assist the young of others, and if the tendency to call is heritable, then those who call will have more surviving, reproducing offspring and become the dominant forms in the population. In this case, the function of calling is to warn vulnerable offspring (e.g., see Sherman, 1977).

As will be discovered later, even though the *concept* of function is simple, the identification of the function of particular communicative acts is often far from straightforward, for several reasons. First, the use of a particular act may serve more than one function, and some of these functions may not even be communicative (e.g., calling by rat pups when isolated from the nest not only induces the mother to retrieve the caller, but also is part of a physiological mechanism to keep the pup warm – see Chapter 4). Second, a signal may have effects other than its function(s); for example, territorial male bicolor damselfish court females with chirping sounds, but their chirps also have the effect of eliciting competitive courtship activity by neighboring territorial males (Myrberg, 1981). A major source of such nonfunctional effects is the active assessment activities of nontarget individuals. These nonfunctional effects can, like functional effects, influence signal form through natural selection.

1.1.2 Sexual selection

In 1871, Darwin identified what he considered to be a different kind of selection, and a distinct category of functions of behavior, in his book *The Descent of Man, and Selection in Relation to Sex* (reprinted in 1981). He proposed that some signals by males, particularly those like the peacock's tail that appeared to be of no utility in enhancing survival, must have been favored by selection due to female esthetic preference, which he called sexual selection. These, he surmised, enhanced a male's reproductive success, not by better adapting him to ecological sources of selection such as predation, but by increasing his attractiveness to females: 'The females are most excited by, or prefer pairing with, the more ornamented males, or those which are the best songsters, or play the best antics.' (Darwin, 1871/1981, i, p. 262). Darwin also felt that such visual signals may have spread because they facilitated the efforts of males to compete with other males for sexual access to females, perhaps by intimidating competing males (e.g., Darwin, 1871/1981, i, p. 279). We will see later that this hypothesis, that the esthetic/emotional effects of signals on others could be a source of selection on the form of signals, is of significant importance to current work on animal vocal communication.

Intriguingly, Darwin seemed skeptical that vocalizations were of use in male–male sexual competition, even though he emphasized their esthetic impact on females (e.g., Darwin, 1871/1981, ii, pp. 275–6; see also Fig. 1.1). Darwin was uncharacteristically off the mark in his interpretations of the significance of vocalizations: 'The cause of widely different sounds being uttered under different emotions and sensations is a very obscure subject. Nor does the rule always hold good that there is any marked difference. For instance with the dog, the bark of anger and that of joy do not differ much, though they can be distinguished. It is not probable that any precise explanation of the cause or source of each particular sound, under different states of the mind, will ever be given.' (Darwin, 1872/1965, pp. 85–6)

Of course, Darwin may have been limited in his opportunity to detect variation in vocal structure and to understand the functions of vocalizations because there were no means in the late 1800s of objectively measuring sound structure (such as tape recorders and sound-analysis systems; see below). In contrast, Darwin's ability to study the evolution and communicative utility of visual signals was doubtlessly facilitated by the availability of still photography and drawings as means of describing their structure. As a result of these technological limitations in acoustics, visual signals dominated communication research well into the twentieth century.

1.1.3 Evolutionary history of signals

It was one thing to describe a signal and identify its current function, but a different matter to identify its evolutionary origins. Darwin's answer, in *The Expression of the Emotions in Man and Animals* (Darwin, 1872/1965), was that many signals originated as results of other processes, and not because of their utility in communication. The vocalizations of air-breathing vertebrates, for example, may have originated as a byproduct of the violent muscular contractions of the respiratory system that come with strong excitement, resulting in 'purposeless sounds' (Darwin, 1871/1981, ii, p. 331).

Although Darwin felt that most emotional expressions originated for reasons other than communication, he was clearly aware of the importance of those expressions in communication (Darwin, 1872/1965, p. 364). Thus actions originating for one reason can subsequently acquire communicative function. Some of Darwin's ideas about how noncommunicative actions became signals are not accepted today; he believed that habitual use of expressions could induce heritable variation, that is, result in the habit's transmission to offspring. The view that acquired characters could be inherited came to be labeled Lamarckian, and widely rejected as a likely mechanism.

An example of a more modern, and also quite Darwinian explanation of this process is as follows (Ploog, 1992; Andersson, 1994). In ancestral aquatic vertebrates, the larynx (which ultimately became the vocal apparatus) was essentially a valve in the floor of the pharynx (the muscular tube connecting the mouth with the esophagus). While the individual was under water, this valve served to cap the swim bladder (which was the evolutionary precursor to the lungs). At the water's surface, the valve opened, allowing air to be forced into the swim bladder, which provided buoyancy. During the evolutionary transition from aquatic to terrestrial life, the swim bladder acquired respiratory function. Associated with this change in function of the swim bladder was an elaboration of the larynx;

this included the evolution of fibers to pull the larynx open, and subsequently the addition of cartilage, both changes allowing more air to be pulled into the lungs. This elaboration of the larynx for its respiratory function created a structure that tended to produce sound, for example, when air is abruptly inhaled when the individual is startled. Due to some effect of such incidental sounds on conspecifics, the sound producer may have been more successful in its social interactions and therefore more reproductively successful; as a result, the tendency to produce such sounds during social interactions would spread through natural or sexual selection. In this way, sounds produced as byproducts of respiration may subsequently become signals as a consequence of selection arising from their social effects. Females might respond positively to such expressive males for several reasons; for example, the loudness or persistence of sounds might be used as a cue to assess a potential mate's stamina, thereby making it possible for the female to choose the best sire for her offspring. Under these conditions, emotional expressions could become more widespread with each successive generation because they enhanced the male's reproductive success by increasing his attractiveness to females.

Once a signal originates within a species, it may change, remain the same over time, or disappear (Moynihan, 1970). Furthermore, the signal may remain the same or change in new species that arise from old. When a signal of a given species is examined today, how does one decide whether this is a new form or one inherited from an ancestral species? Clearly, a dilemma is that such questions deal with changes that are too slow for humans to study directly. So, we create indirect methods, based on the study of the effects of those slow changes. Darwin's answer was to seek the same signal in other species; the more widespread the signal, the more likely it is that it was inherited from ancestral species rather than a new creation (see Mayr, 1988, pp. 276–80). Darwin's argument about the evolutionary origins of signals was supported with extensive documentation of the similarities in ways of expressing emotions among many species of animals, including humans. Such use of comparisons among species (now called the comparative method) has become a powerful method for identifying the behavioral 'raw material' from which signals may have been derived, and the nature of the evolutionary changes that have occurred (Tinbergen, 1952).

1.1.4 Ethology

An evolutionary approach to the scientific study of animal behavior, called *ethology*, emerged from the framework provided by Darwin. Ethologists identified the four different categories of questions about behavior that form a part of the organization of this chapter (Tinbergen, 1963). Two of these were the evolutionary questions that have just been discussed in the context of Darwin's work, i.e., about the functions and evolutionary origins of behavior. These *ultimate* questions address *across-generation* influences on behavior. The other two types of questions deal with the influences on behavior that have their effects during the lifetimes of individual organisms – questions about the immediate causes of behavior, and how the capacity to engage in that behavior develops as the animal matures. These latter two questions are discussed later under the heading 'Proximate questions.'

Ethologists often focused on displays, that is, acts specialized evolutionarily for signaling (Fig. 1.1). The Austrian biologist Konrad Lorenz is widely recognized as one of the founders of ethology. Those who followed his approach to the evolution of behavior emphasized the use of displays as traits for the study of evolutionary relationships among closely related species. Such research provides insight more into the evolutionary origin and modification of displays than into the functions of signaling. If Lorenz had studied the Carolina wren described in the Prologue, one of his primary interests would have been in naming, describing, and listing the various vocal displays. He would have compared the number of these calls shared by other species of wrens as a means of inferring how closely related they are, in other words, their phylogenetic history. Lorenz would have been deeply satisfied to have identified the Carolina wren's closest living relative on the basis of the number of shared vocal displays. He would not have concerned himself much with variation in form within each display category (such as the increase in pitch of the female's *chirts* after movement by the hawk she was monitoring). In fact, the variations would have been considered unimportant 'noise' that should be eliminated in order to facilitate between-species comparisons. Later, how this emphasis on the stereotypy of displays slowed discovery of the importance of within-species variation in display structure is discussed.

The Dutch biologist Nikolaas Tinbergen shared with Lorenz the status of a founder of ethology. They approached the subject similarly in some ways. For example, both would have begun a study of Carolina wrens by cataloguing their vocal displays, and both would have compared their results with those for other wren species as a means of discovering how these vocalizations evolved. Tinbergen extended and clarified work on the evolutionary process of creating signals, which was called *ritualization* (Huxley, 1923). According to Tinbergen, ritualization is 'adaptive evolutionary change in the direction of increased efficiency as a signal' (Tinbergen, 1959, p. 41). He noted that signals were often ritualized modifications of some evolutionarily earlier activity. The human smile, for example, involves retraction of the corners of the mouth and exposure of the teeth (see Fig. P.2 in the Prologue; Ohala, 1980). Smiles are derived from a facial expression that originally served to shorten the vocal tract to produce a high-pitched, appeasing vocalization (for example, say 'eee' and you will experience the same association between facial expression and sound structure). Humans have since deleted the vocalization, retaining only the facial expression. This example shows how knowledge of the evolutionary origin of a signal helps to explain its present form. Exposure of the teeth might seem aggressive, a paradoxical feature of a friendly signal, but not in light of this insight.

Tinbergen differed from Lorenz in his emphasis of questions about the adaptive functions of displays. Tinbergen and his students made extensive use of 'models' of natural stimuli to study their effects. For example, Tinbergen used fish models varying in the amount of red on their bellies to see if the naturally red belly of stickleback males was a signal. If Tinbergen were involved today in the study of vocal communication, for example of the sort described between Anna and her mother in the Prologue, he would use the technique of playing back recordings of vocalizations to study their communicative significance. One detail that he would note is that Anna ceased to cry when comforted by her mother, but he would also realize that this may have been caused by the sounds her mother made, or by the rocking, or cuddling, or all three. The effects of her mother's voice alone on Anna could be assessed by playing back tape recordings of those sounds, without the rocking or cuddling (Box 1.1).

Tinbergen's interest in adaptation led to examination of the significance of the structure of signals. For example, he was struck by the widespread similarity in the threat and appeasement postures of birds, and speculated on its sources. The bill is typically involved in threat, he noted; perhaps threat postures were similar in many species because they were all evolutionarily derived from preparations to use this organ of aggression. But why are appeasement postures similar? Perhaps, he hypothesized, natural selection favors the removal of stimuli during appeasement that might

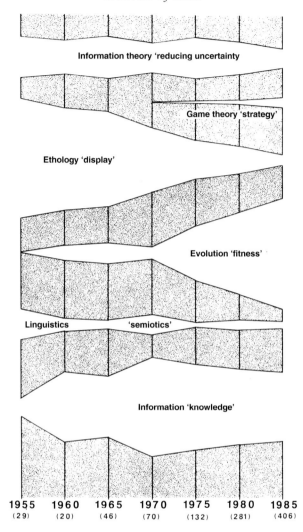

Information theory 'reducing uncertainty

Game theory 'strategy'

Ethology 'display'

Evolution 'fitness'

Linguistics   'semiotics'

Information 'knowledge'

1955	1960	1965	1970	1975	1980	1985
(29)	(20)	(46)	(70)	(132)	(281)	(406)

Fig. 1.1. A short history of animal communication biology. Each concept is represented over time in proportion to the width of its space. The numbers of research papers published using each concept are summed over five-year intervals. Only three of the six concepts recognized today were prevalent in 1950: information, ethology, and linguistics. Why is 'evolution' not prevalent from the beginning? In the early 1950s, evolutionary biology was preoccupied with species-level questions (e.g., species limits, species distinctiveness, isolating mechanisms). The neo-Darwinian 'modern synthesis' (Mayr, 1963) brought an exciting new insight to our perception of the process of speciation by merging formerly divergent theories in genetics and natural selection. Ethologists also viewed their subject, including animal communication, in the new light of the 'modern synthesis.' They were overwhelmingly concerned with animal communication at the species level;

evoke aggression. This might be accomplished by doing the opposite of what is done during threat. There are only a limited number of ways of doing the opposite of displaying the bill, such as turning it away from the adversary, either horizontally, or vertically up or down, or withdrawing the bill. These limited options for 'hiding' the bill may have been responsible for cross-species similarities in appeasement displays.

Fig. 1.1 (*cont.*)
they had little interest in intraspecific, intrasexual, or individual levels (Dawkins, 1976). This historical accident, the invention of field tape recorders and spectrographs at the same time as the biological species concept was formulated, led to the species-level emphasis. Because of this emphasis, the classic ethological generalizations were not often predictive of behavior in individual behaving animals (Moynihan, 1973). Classic ethology had not produced a focus that was deemed useful in the new evolutionary views of the 1970s (Dawkins & Krebs, 1978; Hinde, 1981).

The six concepts are distinguished as follows. Information takes an everyday dictionary definition of *information* (i.e., 'knowledge'), whereas information theory is a logical mathematical construct with *information* defined as that which reduces 'uncertainty.' Together, the *information* concepts constitute over 50 percent of the communication studies since 1955. The information theory concept derives from the mathematical theory of communication as presented by Shannon and Weaver (1949). The rigorous logic in information theory has been, and still is, attractive to many biologists (e.g., Wilson, 1975; Hailman, 1977; Wiley & Richards, 1982). Most researchers do not use the term *information* in the sense meant by information theory, but W. John Smith has synthesized the two ideas (1977).

Ethology, as mentioned above, considered communication at the species level, whereas the evolution concept places the individual's fitness foremost – linked, ultimately, to population genetics. An example of the ethological view is that birdsong has evolved as a mechanism to avoid hybridization between species. Support comes from the fact that most birdsongs are 'species distinctive'. An evolutionary approach asks how species distinctiveness benefits the *individuals* that exhibit it. As explained in the text, the answer does not take the origin and maintenance of any behavior for granted as simply typical of a species. For example, species distinctiveness could be an inevitable consequence of intraspecific competition and not a result of interspecific competition. The classic ethological emphases declined rapidly from 1970 on, replaced by the new evolutionary emphasis on individual and genic levels of selection.

The linguistic concept is identified by terminology such as zoosemiotics (Sebeok, 1967) and pragmatic, syntactic, semantic levels as mentioned above. As one of the early concepts, it became linked more strongly with *information* in the late 1950s. Game theory became more popular in 1975. Its assumptions fit well with the new evolutionary view. Like information theory, game theory has a statistical logic but coupled with the logic of natural selection. It is easily identified in papers by terms such as game, winners and losers, asymmetries, strategy, and ESS (evolutionarily stable strategy – a different concept from the ESS of Chapter 3).

Box 1.1. *The utility and limitations of playback methods.*

In the context of acoustic communication, *playbacks* involve playing a
recording of an acoustic signal and noting its effects on one or more
individuals. The most important feature of playbacks is that they are a
form of controlled experiment. Controlled experiments make it possible to
distinguish between correlational and causal relationships. For example, it
was noted in the Prologue that Anna ceased to cry when comforted by
her mother. That cessation of crying was correlated with at least three
features of the mother's behavior: the stimulation provided when the
mother made soothing sounds, rocked Anna, and cuddled her. However,
we cannot judge which of these three actually caused Anna to stop crying
unless a controlled experiment is performed in which each one of the
three is varied separately while holding the other two constant. The
effects on Anna of mother's soothing voice alone could be assessed by
playing back tape recordings of such sounds to Anna, without the rocking
or cuddling. Anna's behavior under these sound conditions could then be
compared with her behavior in the presence of other *control* sounds, again
without the rocking and cuddling. The purpose of control sounds is to
assess the extent to which the effects of sound are dependent upon specific
properties of the sounds, such as the distinctive gentle shushing and quiet
voice with long, smooth, low and falling pitch contours used by Anna's
mother. In this case, appropriate control sounds would be recordings of
Anna's mother in other contexts, for example, speaking with her adult
friend. Although this precise playback experiment has not been done,
other studies of the effects of sound stimulation have demonstrated that
at least three of the features of mother's soothing voice are exceptionally
effective in soothing distressed infants, even in the absence of actual
spoken language, rocking, and cuddling (Fernald, 1992): low frequency
(versus high frequency: Bench, 1969); continuous (versus intermittent:
Birns *et al.*, 1965); and broad band (versus the sound of a human voice:
Watterson & Riccillo, 1983).

It is important to keep in mind the extent to which individuals
responding to playbacks engage in active assessment. It was noted above,
for example, that playback experiments remove many stimuli that
normally accompany vocalizations. Nevertheless, the actively assessing
subjects of the playbacks may respond to more than just the structure of
the song played to them. The apparent behavior of the 'singer' may also
be inferred (Pepperberg, 1992). For example, songs played back in the
middle of a bird's territory appear to be interpreted as a territorial
intrusion. However, such simulated 'intruders' typically continue to sing
in the same way, with no modification in location or patterning of singing
when the resident approaches, displays aggressively, and countersings.
This is abnormal behavior for an actual intruder, and subjects of
playbacks may treat them as such (Pepperberg, 1992).

The significance of such contextual factors in playback studies depends upon the normal circumstances of the vocalizations under study. In some circumstances, opportunities to interact with the vocalizer or make use of contextual cues are quite limited. These conditions often hold, for example, when an individual first detects a predator and emits an alarm call. The urgency of this situation forces those who hear the call to base their early reactions on its structure, emphasizing reaching refuge and surveying for danger, rather than focusing on the caller (see, for example, Leger & Owings, 1978). In other circumstances, such as the birdsong playbacks discussed above, dealing with the vocalizer is the primary concern of those who hear the vocalization. Here, traditional playback methods provide less-valid simulations of natural circumstances; vocal structure is precisely duplicated but vocal behavior is not.

Simulation of natural singing behavior is difficult under the technical restrictions of traditional playback equipment. For example, imagine an experimenter who wanted to make birdsong playbacks more realistic. She or he might want to be able to switch to one of several different playback sounds, in an immediate, interactive way, depending on the subject's response to the initial playback. Given the time required to change audiotapes on a recorder, each optional sound might require a different playback device. The bulk and expense of the required equipment would quickly limit the feasibility of such methods. However, remarkable progress has been made recently in the development of small but powerful portable computers with software for digital sound manipulation and playback (Dabelsteen, 1992). Many sounds can be stored in such devices and selected for playback quickly enough to permit truly interactive playback experiments. The use of such methods suggests that more realistic, interactive playbacks of birdsong evoke stronger and more discriminative responses (Dabelsteen & Pedersen, 1990). These technical advances have set the stage for systematic studies of the effects of both the structure of vocalizations and the interactive contexts in which the vocalizations occur.

1.1.5 Self-interest: the logic of natural selection

From the perspective of natural selection, whose interests should be served by the activity of vocalizing? Prior to the 1960s, no clear distinction was made between benefits that accrue to the individual's group or species, and those accumulating to the individual itself. At that time, it was common to assume that natural selection would produce individuals who cooperated and even reproduced for the betterment of the species (e.g., Huxley, 1938). This viewpoint, which came to be called *group-selectionist*, was developed most explicitly by Wynne-Edwards (1962), and was systematically critiqued in Williams' well-known book (1966). In the 1960s, Williams and a few other biologists began to point out an implication of a Darwinian approach that not even Darwin understood clearly – natural selection should be as likely to produce conflict as cooperation among the members of a species. In other words, individuals should vocalize in ways that make them more successful reproductively than other members of their group.

The principal defect identified by Williams in the idea of group selection can be illustrated with a brief discussion of the calling that prairie dogs engage in while dealing with mammalian predators. Such behavior superficially has the appearance of self-sacrifice, and thus might be considered a trait shaped by group selection; that is, barking appears to incur costs to the caller (in energy and perhaps increased conspicuousness to the predator), while apparently benefiting others by warning them. One might hypothesize that groups containing barkers have lost fewer members to coyotes, and therefore survived longer than groups in which no one barked; in this hypothetical scenario, the species would come to consist primarily of groups in which everyone called. In order to understand the flaw in such group-selectionist thinking, we must imagine the impact on such groups of natural selection, which should continue operating at the level of individual reproductive success even if group selection is also underway. If a mutant individual were to appear who does not call, it would still receive the benefits of calling by all others, but incur none of the costs. Because a noncaller is free from costs borne by all others, it should contribute more offspring to subsequent generations, and thus become the most frequent type within the population. In the face of natural selection at the level of individuals, groups of self-sacrificers are inherently unstable.

Why, then, does natural selection maintain calling? First, individual natural selection provides a partial answer to this question: the barks warn the caller's offspring in part, and thus foster the caller's production of reproducing offspring (Hoogland, 1983). A second part of the answer lies in the concept of 'kin selection' (Hamilton, 1964): some of the beneficiaries of barking are the caller's nondescendant relatives, such as cousins and siblings (Hoogland, 1983). Facilitating their survival to reproduction also contributes genes like the caller's to the next generation, and is therefore also favored by selection.

The 'bottom line' in natural selection is individual success at fostering the passage of genes like one's own to subsequent generations; those traits that promote such success *relative to other members of the popula-*

tion are the ones that will spread over successive generations. Such a process should favor behavioral tendencies that are in the interests of the actor; natural selection favors a trait because of its 'selfish' effects, not because of the effects of the trait on the group or species as a whole. This realization had a profound impact on thinking about animal behavior. Prior to this change, group-selectionist thinking had led to the expectation that cooperation among members of the same species is the rule, and conflict the exception. Now, cooperation requires special explanations, and within-species conflict of interest becomes the default case. More recently, this Darwinian conflict has been described between different classes of individual, such as male versus female (Trivers, 1972), and parent versus offspring (Trivers, 1974).

Burghardt (1970) was among the first to introduce self-interested thinking to the study of communication. Maynard Smith and Price (1973) extended this idea by applying a game-theory approach to an understanding of ritualized fighting, which they defined as the substitution of signals for actual combat. Ritualized fighting can be illustrated with the hooded warblers described in the Prologue. Remember that the males of this species re-established their territories each spring, responding especially vigorosly to males that were not their neighbors from previous years (Godard, 1991). However, the vigor was primarily in display rather than fighting: approach and display of the black throat and drooped wings usually sufficed to repel newcomers.

Maynard Smith and Price wanted to know whether refraining from actual combat was the product of group selection (against injuring conspecifics), or could be explained in terms of individual reproductive advantage. They used a game-theoretic approach, which is a mathematical formulation designed to deal with conflicts of interest, such as those that characterize games like poker and chess, involving competition for limited resources. New insights into the adaptive significance of ritualized fighting resulted from focusing on two different social consequences of aggressive behavior. First, they noted that escalating a fight may be costly to the escalator due to the increased risk of injury from retaliation. So, individuals might refrain from fighting not to avoid hurting others, but to avoid being hurt. Second, they demonstrated how social context can be a major determinant of what kind of behavior works best in dealing with others. Maynard Smith (1982) used a simple mathematical model, in which individuals could behave only as 'hawks' (all-out fighters) or 'doves' (ritualized fighters), to show that the most profitable way of behaving during a conflict depended on what other members of the population were doing. In a population of mostly doves, a hawk would do very well because it would always win and rarely pay the price of an all-out fight. On the other hand, an occasional dove in a predominantly hawk population would do well, because doves would be free of the heavy costs of injury frequently paid by hawks fighting one another. In other words, either hawks or doves could be at an advantage, depending upon which was proportionally rare; however, this advantage declines as that way of behaving becomes proportionally more common. Under such conditions, natural selection would favor a mix of hawks and doves, in the proportions at which the benefit/cost ratios of the two strategies are equal. This mix was called an evolutionarily stable strategy (ESS) because, once established, no alternative strategy among the existing options could supplant it. Thus, substitution of signals for all-out fighting is not necessarily the product of selection for avoiding hurting others; it may instead reflect selection for self-interested efforts to avoid *being* hurt.

This self-interested approach to communication was further stimulated by Dawkins and Krebs' (1978) 'manipulation' view of animal communication. They chose the label manipulation in order to criticize two features of then-current approaches to the study of communication. They were objecting to the idea that communication is founded on the exchange of information (see Box 1.2) and they were challenging the idea that communication is a cooperative process, involving little conflict of interest. Dawkins and Krebs chose the term manipulation for its pragmatic, self-interested connotations. (By pragmatic we mean that communicative behavior must be viewed as being shaped by its consequences and history.) Signals are acts selected for their effectiveness in manipulating the behavior of others to the signaler's advantage.

Box 1.2. *The concept of information in communication.*

Use of the concept of information has spread rapidly since its introduction to animal communication studies in the 1950s (Marler, 1956). *Information* is defined in two different ways, and its attractiveness to researchers is a product of features of both meanings. One definition is derived from 'information theory,' which was developed by communications engineers (Shannon & Weaver, 1949); here, information is defined quantitatively as the 'reduction of uncertainty' provided by a signal (Young, 1954; Wiley, 1983). The other definition deals with knowledge or semantic content, that is, what the information is about

(Smith, 1977). The engineering definition has the appeal of mathematical precision, the semantic definition, the intuitive attractiveness of an everyday idea. The complementary appeals of these two meanings combined to make information a very attractive concept. According to information theory, information transfer can be assessed by relating events in the channel (e.g., the pattern of signaling activity) to those in the receiving animal. The result is a quantitative measure of information, specified in *bits*. A bit is the quantity of information needed to reduce the receiver's uncertainty by half. Miller (1965, p. 194) illustrated this idea in the following way. 'Legend says the American Revolution was begun by a signal to Paul Revere from Old North Church steeple. It could have been one or two lights – "one if by land or two if by sea." If the alternatives were equally probable, the signal conveyed only one bit of information, resolving the uncertainty in a binary choice.'

Semantic information deals with the *significance* of signals. The one bit of information conveyed to Paul Revere carried a vast amount of meaning whose magnitude must be assessed in terms other than bits. For example, the same lights could be used again to convey one bit of information, this time about whether tea was to be served at 4:00 or 5:00 p.m. The tea and British-attack information would be similar, quantitatively, but vastly different in terms of their significance for the lives of the colonists. The concept of semantic information is derived primarily from semiotics, the theory of signs (Peirce, 1958; Cherry, 1966). Using a semiotic approach, Smith (1977, p. 193) defined information as 'an abstract property of entities and events that makes their characteristics predictable to individuals . . . (It) enables . . . individuals to make choices, to select their activities . . . appropriately for their needs and opportunities.' Critical features of semantic information include its *utility* for making decisions, the *importance* of the selected actions, and the use of *messages* rather than bits as basic units. Ultimately, the transmission of information came to be viewed as a major causal factor and central defining property of communication, as the following quotes illustrate.

Communication consists of the transmission of information from one animal to another. Information is encoded by one individual into a signal. When received by another animal, this information undergoes decoding, while still retaining a specifiable relationship to the encoded information
(Green & Marler, 1979, p. 73).

Animals need to be informed. Obtaining information is crucial to their ability to respond actively to their surroundings . . . Selection favors individuals able to obtain and use such information from each other . . . This is only part of the story, however. Individuals can benefit not just from obtaining information from each other, but also from making it available to each other. There are evolutionary pressures upon animals to become more useful as sources of information . . . ; it is adaptive to provide

information that may lead a recipient to behave appropriately to the
communicator's needs

(Smith, 1977, p. 9).

Signals cause changes in the behavior of other animals because of their
information content rather than physical force, and have been specifically
altered by selection in some way to achieve that end . . .

(Guilford & Dawkins, 1992, p.384)

1.1.6 Selfish perceivers

At approximately the same time that Dawkins and Krebs were formulat-
ing their manipulation view, Zahavi (1977) was developing his handicap
hypothesis to account for reliability in communicative systems. Relative
to Dawkins and Krebs, Zahavi emphasized the importance of the other
side of communication – the receiver side. Aware of the advantages to
signalers of manipulative use of signals, he raised the question of what
limited such manipulation. His answer was that receivers are selected to
make use only of reliable information sources. Signals were more likely to
be reliable if they were costly: '. . . a signal is reliable when the difficulty
of its performance is related to its meaning in quantity and quality. . . A
signal which means "I am very strong" should be more difficult to send
than the signal "I am strong". Furthermore, a stronger individual should
find it less costly to signal its strength than a weaker one, cost being
measured in terms of reproductive potential.' (Zahavi, 1977).
Therefore, receivers are selected to be responsive only to costly signals.
The result, Zahavi argued, should be an imposition of cost on signalers
by receivers, which was said to 'handicap' signalers. He illustrated his
point with work on roaring by red deer stags (Clutton-Brock & Albon,
1979), which is used during male–male competition. Roaring remained a
reliable index of male condition, Zahavi proposed, because producing
these vocalizations was energetically very expensive. Another way to
put this is that roaring 'overdraws' the energy accounts of both strong
and weak males, but it only bankrupts the weak males.
    Emphasis on the perceiver role provided a rebuttal to the manipulation
view. Rampant manipulation was not likely because individuals are
selected to play the perceiver role in their own interest. Although few
at that time subscribed to the handicap hypothesis, many subsequently
echoed a more general form of the rebuttal that follows from Zahavi's

thinking, that is, that the potential for signalers to manipulate perceivers is constrained by the perceivers' pursuit of their own interest (Hennessy *et al.*, 1981; Morton, 1982; Cheney & Seyfarth, 1985; Markl, 1985; Smith, 1986b). Table 1.1 lists some concepts from the communication literature that illustrate the growth of interest in assessment.

1.1.7 *Assessment/management: combining selfish receivers and signalers*

Approaches that offer a synthesis of the Dawkins/Krebs and Zahavi viewpoints began to emerge in the early 1980s (Hennessy *et al.*, 1981; Morton, 1982). In this synthesis, communicative systems are thought of as emerging from the dynamic interplay between two equally active roles – one has variously been labeled the manager, manipulator, signaler, or sender role; the other the assessor, mind-reader, perceiver, or receiver role. All individuals play both types of roles in communication, and the logic of natural selection indicates that individuals should be selected for their effectiveness in both. As managers, they should be able to use communication to achieve fitness-enhancing ends by influencing the behavior of others, in part by exploiting their assessment systems. As assessors, they should be capable of making adaptive behavioral adjustments through selective attention to the most reliable cues available for appraising individuals and situations, whether or not these cues arise from signals (Krebs & Dawkins, 1984; Markl, 1985).

Geist's (1974) work on the evolution of horns and fighting strategies foreshadowed the development of this coevolutionary approach. He proposed that agonistic behavior was best understood in terms of an evolutionary interplay of offensive and defensive roles. Geist provided evidence that an individual which hesitates to attack an adversary may be acting in its own interests. An animal with well-developed weapons of aggression is likely to face similar weapons, and therefore a serious risk of retaliation. Furthermore, natural selection arising from such weapons would favor effective defense; as a result, escalation to injurious levels would also be limited by the fact that actual efforts to attack would be parried or blocked by morphological and behavioral defenses.

Morton's (1977; 1982) hypothesis about the origin of motivation-structural rules in acoustic communication illustrates how the evolutionary interplay of management and assessment roles produces communicative systems (Chapter 3). This hypothesis helps us to understand why low-pitched, growl-like sounds, for example, are associated with aggressive motivation in a variety of vertebrate taxa.

Table 1.1. *Evidence of greater interest in the active nature of assessment.*

---

1. *The assessor strategy* (Parker, 1974, p. 224): '. . . any mutant individual able to assess from the conventional fighting stage how its own RHP (resource holding power) compares with that of its opponent would have a selective advantage, since it could withdraw without damage when the RHP of its opponent exceeds its own by a sufficiently large amount.'

2. *Active perceivers* (Hennessy *et al.*, 1981, p. 190): 'In the management metaphor, we retain Dawkins and Krebs' "selfish signaler" view, but emphasize more heavily the impact of perceivers...Both signaler and perceiver are active in communication, but they face different problems. The percipient needs to extract information in terms of what to do, and where and when to do it. In contrast, the signaler needs to obtain a particular kind of performance from the targets.'

3. *Signal interception* (Myrberg, 1981, p. 399): 'Numerous cases exist where information about an individual (the initiator) reaches another for whom that information was not apparently intended. The receiver, if it reacts, does so for its own benefit and frequently the initiator of the interaction is disadvantaged.'

4. *Skeptical perceivers* (Moynihan, 1982, pp. 9–10): 'The world is a difficult and dangerous place for most animals. As a mere precaution, if for no other reason, animals should try to "read" any potentially useful information that is proffered to them. . . (However,) they must assess the information proffered . . . with some considerable degree of scepticism. . .They probably use all sources of information perceivable, not only displays. . . It may be supposed that they test the reliability of each and every bit of news by comparison with other bits (sic) in order to determine if all the bits are consistent with one another.'

5. *Ranging* (Morton, 1982, p. 200): '. . . the goals of singers and listeners differ; there is no selection favoring sharing of information, and singers and listeners are usually at odds with one another . . . selection on listeners favors their ability to determine, as accurately as possible, if singers are truly encroaching on the listener's territory.'

6. *Mind reading* (Krebs & Dawkins, 1984, p. 401): 'The evolution of ritualized signal movements or structures from their precursors is the product of coevolution between roles. We have termed the roles "manipulator" and "mind reader." The manipulator role is selected to alter the behavior of others to its advantage, the mind-reader role to anticipate the future behavior of others.'

6. *Probing* (Owings & Hennessy, 1984, p. 190): 'The concept of feedback raises the possibility of an even more interactive form of information extraction, which might be called "probing." Adult male black-tailed prairie dogs appear to distinguish between residents and nonresidents of their coterie at some distance by rushing at the individual in question, whose reaction aids the male in identification: nonresidents usually flee, but residents crouch, wag their tails, and open their mouths in apparent anticipation of kissing. . . We propose that many social inputs function in part to elicit behavior on which to base further target assessment.'

7. *Sensory exploitation* (Ryan & Rand, 1990, p. 305): 'Some of the most elaborate morphologies and behaviors in the animal kingdom are male traits that function in the attraction and courtship of females. Darwin...suggested that these traits, often evolved because of female preference motivated by an aesthetic sense...(His) notion suggests an important, experimentally verifiable, and usually neglected

Table 1.1. (*cont.*)

approach to the evolution of male traits under sexual selection. As "beauty is in the eye of the beholder," the various properties of the female's sensory system determine which traits are 'pleasing' to its eyes, ears, or nares, or, more precisely, which traits will be favored by sexual selection due to their superior abilities to attract females.'

8. *Receiver psychology* (Guilford & Dawkins, 1991, p. 1): 'It is argued that an important but neglected evolutionary force on animal signals is therefore the psychology of the signal receiver, and that three aspects of receiver psychology (what a receiver finds easy to detect, easy to discriminate, and easy to remember) constitute powerful selective forces in signal design.'

(Remember, for example, the Carolina wren in the Prologue which uttered growls and harsh, low *rasps* while attacking an intruder, in contrast to the high-pitched *pee pee pee* calls uttered by the appeasing intruder.) This relationship between sound structure and motivational state reflects at least three steps in the coevolution of management and assessment. (1) On the basis of physical principles alone, larger individuals should emit sounds of lower pitch and be more dangerous to other individuals; that is, larger size both lowers the resonant frequency of the acoustic apparatus and increases the individual's strength (e.g., Davies & Halliday, 1978). (2) Because physical constraints limit the extent to which the body size/sound-pitch relationship can be broken, assessors should be selected to attend to this reliable pitch cue about dangerousness by judging themselves to be at greater risk, on average, when an adversary emits sounds of lower pitch. (3) Within the limits imposed by physical constraints, managers should be selected to exploit this assessment rule by using lower-pitched sounds when aggressively motivated to maximize their apparent size and deter opponents, and by emitting higher-pitched sounds when their motives are nonaggressive. According to this general formulation, the structure of a signal (e.g., pitch) is not arbitrarily linked to its meaning (e.g., signaler's aggressive motivation). The structure of signals must be evaluated in the contexts of the functions of signaling, the cues used for assessment by the signal target, and the constraints that generate the relationships used in assessment.

The approach applied by Morton to the evolution of communicative systems shares features with some other ideas that have been discussed in this chapter, specifically Darwin's concept of sexual selection as well as a game-theoretic approach (see also Zahavi, 1991). In all three, the major sources of selection on the system are from within rather than outside the species. Darwin considered such social sources of selection to be unusual, but nowadays we know them to be common (Cronin, 1991). When selection arises from biotic interactions (among organisms), whether heterospecifics or conspecifics, the potential exists for reciprocal influences; for example, the manager role is a source of selection on the assessor role, and assessment is also a source of selection on management. As a result, these two roles can become locked in a coevolutionary 'spiral' analogous to an 'arms race,' in which each drives the other indefinitely through evolutionary modifications (Dawkins & Krebs, 1979).

1.2 An evolutionary approach: proximate questions

The concept of natural selection deals only with the average fitness consequences of behaving in particular ways. For example, an ecologist would be unlikely to progress far in studies of foraging behavior by attempting to account for each of the many prey captures per day exclusively on the basis of ultimate concepts. An appeal to proximate processes would be needed, as a complement to the ultimate thinking represented by such efforts as optimal foraging theory, which provides a framework for the evolution of eating behavior.

Proximate processes can be defined as influences on behavior that have their effects during the lifetimes of individuals, rather than across generations as ultimate processes do. Sources of proximate effects on behavior include changes in the states of an individual's physiological and psychological mechanisms, and modifications in the environmental circumstances in which an animal finds itself. As in the investigation of foraging, in the study of communication we can more completely explain each episode by focusing on proximate processes in addition to ultimate processes. If a white-crowned sparrow, for example, responds to the sound of a song in one way on one occasion, and in a different way to exactly the same song on another occasion, we need to know what differed between those two occasions that could account for this difference in response. One proximate determinant of such responses is the level of the hormone testosterone in the male's circulatory system (Wingfield & Hahn, 1994). When levels are high, as they are when females are sexually receptive, resident males respond very aggressively to the sound of intruding males singing. When testosterone levels are low, as they are when a male's mate has already laid her eggs, the intensity of the male's response is also low.

Proximate processes operate in many time frames: in the simplest conceptualizations, two are identified. Immediate effects reflect the narrowest frames – these can be illustrated by the instant response exhibited by a territorial male song sparrow when he hears an intruder singing. Similarly, in the Prologue, the sight of Anna standing in her bouncy chair evoked a quick 'NO!' by her mother. Developmental effects reflect the widest proximate frames. Remember in the Prologue, for example, that the songs a fledgling Carolina wren heard influenced the structure of the songs it sang as an adult. This is true of all songbirds studied to date. We shall see later, in the discussion of tonic communication, that a minimum of three proximate time frames will be needed to account for the complexity of communicative processes.

Recently, more and more scientists have realized that an understanding of proximate processes provides insights into ultimate processes, and vice versa. This can be illustrated by Ryan's (1994) research on sexually selected communication: 'For too long physiologists and ecologists have tended to work in isolation from each other. The dichotomy of proximate versus ultimate sometimes seems not only to justify but also to demand ignorance of the alternative approach (p. 191). . . Female mating preferences are behavioral manifestations of neural properties. As shown in this chapter, these neural properties can be studied, and an understanding of these mechanisms can give us a better appreciation of how these preferences evolve.' (p. 211). As noted in the Prologue, Ryan has demonstrated that female Túngara frogs prefer to mate with males who emit complex calls, containing both *chucks* and *whines*, over males who only *whine*. It is typically assumed that such assessor preferences for particular calls have coevolved with call structure. However, Ryan hypothesized that the sensory bias that accounts for the preference for calls with *chucks* predated chucks and helped generate their evolution. (Ryan's sensory-exploitation hypothesis – chucks evolved as a means of exploiting auditory biases in females.) How might one test this evolutionary hypothesis? One way would be to examine the proximate mechanisms of hearing in close relatives of Túngara frogs, most of whom do not produce chucks. Do their ears exhibit similar response biases? Yes they do; a finding consistent with Ryan's sensory-exploitation hypothesis.

1.2.1 Immediate causation: early ethological concepts

The ethologist Konrad Lorenz proposed that certain motor patterns, such as those used by animals as signals in communication, were distinctive in their proximate causal relationships. Lorenz called these patterns fixed action patterns (FAPs) because they appeared to be quite stereotyped in form. His observations indicated that learning played little role in their development; for this reason, he called the perceptual/motor neural structures that underlay their occurrence and form innate releasing mechanisms (IRMs). Lorenz's thinking on the immediate causation of FAPs was consistent with Darwin's (1872/1965). Lorenz felt that motivational and even emotional factors played a significant role in the production of FAPs; animals sought opportunities to perform FAPs, he asserted, perhaps because their performance is associated with feelings and passions (Lorenz, 1970, p. 325). Although these motor patterns were typically elicited by external stimuli (which he called releasers), they were not like reflexes. Rather, FAPs were internally organized and driven. A releaser was said to trigger a FAP in an individual much as an ignition key unlocks and releases the intrinsic motive potential and patterning of the functioning of an automobile.

FAPs play a dual role in the social life of animals, according to both Lorenz and Tinbergen. They constitute the response to a releaser (signal or display) from another member of one's own species (a conspecific); at the same time, they serve as the releaser for the conspecific's next response in an ongoing interaction. So, the conception of communication in its early ethological form was of the performance of a relatively invariant ritual, with displays by each participant serving as 'keys' that unlocked and released a response-display in the other, which in turn released the next display in the first, and so on. Displays and the IRMs of the targets of displays were said to be shaped by natural selection to fit together much as a key and lock are. The properties of displays and innate releasing mechanisms were said to have evolved together, as a communicative unit, favored by natural selection because it facilitated orderly interactions among the members of a species.

1.2.2 The importance of variation in signal structure: Karl von Frisch

The ethologist Karl von Frisch emphasized proximate processes more than Tinbergen and Lorenz did. Nevertheless, he reflected an evolutionary approach in his exploration of mechanisms evolved to deal with the distinctive ecological demands faced by a species. Von Frisch is best known for his discovery of a remarkable communicative system that honeybees use in foraging. Honeybees live in hives composed of tens of thousands of individuals, most of whom are worker females who gather pollen and nectar over a wide area. These workers adjust their foraging activity to the changing availability and nutritional value of food sources, usually concentrating on the richest sites available within reasonable flying distance. These coordinated adjustments in foraging are made possible by communication. When a honeybee finds a food source, it returns to the hive and performs what von Frisch called a dance. During 20 years of careful experimentation, von Frisch discovered that the tempo and orientation of this dance vary substantially from one occasion to another, and that this variation correlates with the distance, direction, and richness of the food source. Other workers in the hive follow the dancer closely, picking up odors and information about direction and distance, and subsequently use this information to search for food.

Von Frisch's discovery of the importance of variation in signal structure foreshadowed significant developments in the modern study of animal communication. His approach would have been difficult to apply to vocalizations, because of the absence of methods for assessing acoustic variation, as discussed earlier. However, with the advancement of audio technology, the gate was opened for similar advances in the study of vocal communication.

1.2.3 The importance of variation in signal structure: audio technology

About 1950, auditory signals began to replace visual ones as the most commonly studied form of communication. This rapid expansion of bioacoustic studies was due almost entirely to the ease with which sounds could be recorded, analyzed, and used as playback 'models.' None of the technological advances (battery-operated tape recorders, for example) was developed specifically for the study of animal auditory communication. Nevertheless, the implications for this field were remarkable; for example, only one and a half decades ago, experimental tapes were constructed by cutting and splicing recording tape. Now, computers can be used to store and play back sounds, and even to modify or synthesize them. Playing back vocalizations to animals, first performed in 1891 on monkeys in zoos using the graphophone (Garner, 1892), became a standard experimental technique (see Box 1.1). Thanks to our human reliance on vision and hearing, the use of technology developed for human electronic communication and entertainment continues to provide new devices for communication studies (Baptista & Gaunt, 1994).

These technical developments began to alleviate the limitations that hampered Darwin's insight into vocal signals. Especially important was the invention of the sound spectrograph, a device for objectively describing the structure of sounds. First applied to animal vocalizations during the 1950s, the spectrograph set the stage for major advances in the study of vocal communication by making possible the first systematic studies of the relationship between the form and function of vocalizations. Peter Marler (1955) was among the first biologists to break this new scientific ground.

Consistent with the ethological view that animal signals were stereotyped, sound spectrographs were initially used to provide an objective description of a 'typical' example of each type of call in a species' repertoire. Nevertheless, this descriptive tool made it possible to detect within-type structural variation in calls not conspicuous to the human auditory system. Once detected, this variation could be explored to determine whether it reflected 'errors' in call production, or whether variation within the same call type made a difference in communication. Colin Beer (1973; 1975) pioneered in such work, discovering evidence that adult laughing gulls varied the structure of their long-calls depending on whether their fledgling chicks or other adults were the targets of these vocalizations (Fig. 1.2). The availability of playback methods made it possible to verify experimentally that different variants of calls make a communicative difference. Indeed, it was the use of playbacks that led Beer to his discovery of the structural variation in long-calls.

1.2.4 Importation of concepts developed for the study of human communication: the proximate mechanisms of animal communication may be more complex than previously believed

In 1956, Peter Marler published his paper 'The voice of the chaffinch and its function as language.' He used spectrograms to describe the vocal repertoire of this songbird, and proposed that information transfer is the proximate process whereby displays affect the behavior of others. This approach to the study of communication was founded on an amalgamation of information theory and semiotics (see Box 1.2), conceptual frameworks created to facilitate the understanding of human communication. Marler's appeal to the process of information exchange repre-

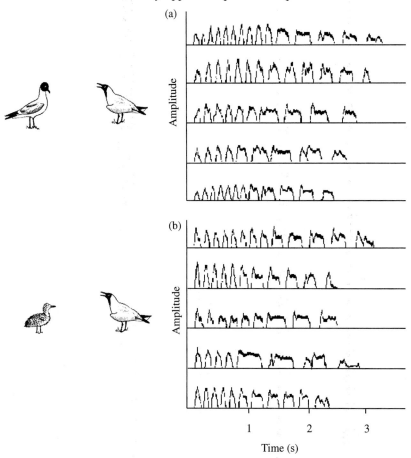

Fig. 1.2. Variation in call structure within a type of call, the *long-call* of laughing gulls. This illustrates one of the pioneering discoveries of differences in the structure of a particular type of call, depending on the context in which it is emitted. In this case, the difference is in patterns of amplitude modulation. (a) Long-calls of five different gulls, each directed at another adult. The first two notes are softer than the rest, or there is a progressive increase in amplitude across the first several notes. (b) Long-calls of the same five gulls, this time directed at the callers' offspring. In contrast with adult-directed calls, the first two notes are of higher amplitude or equal to subsequent notes. (Spectrograms courtesy of Colin Beer.)

sented a significant departure from the more mechanical ethological releaser concept discussed above. As noted by Smith (1977), this new approach was more effective in accounting for the variation in communicative behavior that was becoming increasingly evident.

The overview of semiotics provided by Cherry (1957) was an important resource for both Peter Marler (1961) and W. John Smith (1963). Semiotics is the theory of signs, and signs are symbols for something else. The word coyote, for example, represents a species of dog-like carnivore. A semiotic approach to animal communication treats animal signals as representing, or making information available about, something else (e.g., the bark vocalizations of vervet monkey's may represent leopards – Cheney & Seyfarth, 1990). This approach identifies three perspectives from which communication can be studied. *Syntactics* deals with the structure of communicative systems, but not the real-world things that the system functions to deal with. Signals are analyzed as physical entities, apart from their function, and rules are identified for combining signals. When applied to human language, syntactics would deal with such things as the structure of words and the grammatical rules for combining words into sentences. The *semantic* level takes one step toward relating communicative systems to real-world matters by studying what a signal refers to. In human language, questions of semantics are being raised when one seeks the definition of a word in the dictionary. According to Smith (1977), one studies semantics by identifying the 'messages' of signals; messages are those things/properties with which emission of a signal is related. For example, a call that consistently occurred just before an animal attacked would be said to contain the message that the caller is ready to attack. Note that questions of function are not part of semantics; when one asks what the function of emitting a signal is, one is dealing with *pragmatics*, the most inclusive of the three semiotic levels. The pragmatic level encompasses signals, their messages, and the uses to which signals are put by individuals participating in communication.

The application of such conceptual frameworks to the study of animal communication generated a major shift in the kinds of questions that were posed. Rather than focusing their proximate questions on the causation and ontogeny of signaling behavior, as early ethologists did, more modern animal communication researchers began to emphasize questions about information – what information signals make available (semantics) and transmit to perceivers (pragmatics), and how animals develop the ability to provide and extract that information. Departure from the ethological releaser concept began to raise the possibility that animal communication is more complex than previously believed (e.g., more human language-like).

The recognition of functional variation in vocal structure (mentioned above) also fed growth in appreciation of the complexity of animal communication. Animal vocal repertoires were generally believed to be quite limited (Moynihan, 1970). But attention to functional variation suggested much higher potential for complexity than provided by the stereotyped-display concept (see Fig. 1.2; Marler, 1955).

Even more complexity was suggested by studies of animal calls that appeared to be loosely similar to some nouns of human speech. For example, vervet monkeys 'bark' primarily when they see a leopard; play-backs of barks elicit behavior very similar to that evoked by the sight of a leopard, suggesting that barks function as labels for leopards (Seyfarth, Cheney & Marler, 1980). In semiotic terms, one would say that the highly specific message of barks is transmitted to perceivers.

Nevertheless, the behavior of perceivers in the presence of barks depends on much more than the message 'leopard,' as was demonstrated by the following study. If the barks of a particular individual are played back repeatedly in the absence of leopards, perceivers cease to evince concern at those barks even though their reactions to the barks of other vervets remain unchanged; that is, they behave as though the barker is unreliable because it 'cried wolf' too many times (Cheney & Seyfarth, 1988).

Observations of this sort further demonstrated the limitations of the releaser concept. An additional idea – context – was required. Smith (1965; 1991b) suggested that animals can enrich the information extractable from a signal with that available from context. He identified two categories of contextual information – (1) inputs concurrent with the signal, and (2) the perceiver's memory of past experiences with that signal (as in the example of crying wolf) – and proposed that the meaning of signals to perceivers depends upon both message and contextual information. Such demonstrations of the multiple determinants of responses to signals reinforced the quest for similarities between human and nonhuman communication, and provided further evidence that animal communication is more complex than it was believed to be during the middle of the twentieth century.

Smith predicted that the use of contextual information by receivers during communication would be widespread. He based his prediction in part on the belief current at that time that display repertoires were small. Small repertoires, Smith felt, were incapable of accounting for the complexity of animal communication without the support of contextual information. Ironically, Smith's prediction of the importance of context was correct, but for the wrong reason. Context is important because of the active nature of the assessment role, not because signal repertoires are so limited (e.g., see Smith, 1986b). Much recent work supports the importance of context (Leger, 1993).

1.2.5 *Resistance to informational thinking in the study of animal communication*

As noted earlier, the study of animal behavior was undergoing a revolution during the 1960s and 1970s as individuals explored the self-interested implications of the logic of natural selection. We have already described how Richard Dawkins and John Krebs (1978) helped bring this revolution to animal communication with their influential article entitled 'Animal signals: Information or manipulation?'. They chose the label manipulation in order to criticize both the idea that communicative systems were mutualistic, and the idea that communication involves the exchange of information. Their approach is illustrated by the following quote. 'We are contrasting two attitudes to the evolution of animal signals. One attitude, which we have here called classical, emphasises cooperation between individuals. Cooperation is facilitated if information is shared. . . The other attitude, which we espouse, emphasises the struggle between individuals. If information is shared at all it is likely to be false information, but it is probably better to abandon the concept of information altogether. Natural selection favors individuals who successfully manipulate the behavior of other individuals, whether or not this is to the advantage of the manipulated individuals.' (p. 309)

How is the behavior of others manipulated, if not through the transfer of information? Dawkins and Krebs proposed analogies such as propaganda, persuasion, and advertising, which emphasize the role of motivation and emotion more than they do the role of information-processing systems. They argued, for example, that 'Advertisements are not there to inform, or to misinform, they are there to *persuade*' (p. 305). Analogously, a courting male is not there to inform, or to misinform, but to persuade a female to mate with him. Perhaps, they propose, this way of thinking provides insight into the remarkable elaboration that can be found in the form of some signals, such as the beautiful songs of some birds, or the stunning structure of a peacock's tail. These may be analogous to the oratorical methods used by effective speakers, to *sway* the audience. Such features of communicative behavior may reflect the effects of natural selection, favoring those signal forms most effective at inducing others to behave in ways beneficial to the signaler.

Reactions to Dawkins and Krebs included both praise of their use of the logic of natural selection, and identification of a problem with their formulation. Myrberg (1981), Beer (1982), Wiley (1983) and Smith (1986a) insisted that 'information' and 'manipulation' are complementary rather than alternative concepts. Information, several argued, has to do with the proximate coupling between signaling and situation (e.g., semantics), or between signaling and the behavior of perceivers (e.g., pragmatics). The term manipulation was used metaphorically and is actually defined in ultimate terms, i.e., of the average impact of the signal on signaler and perceiver fitness. The proximate mechanism even for manipulation, several argued, is the transmission or withholding of information. Little note was made of Dawkins and Krebs' emphasis of motivational and emotional mechanisms, rather than information-processing systems. Such emphasis of information-processing, and de-emphasis of emotion and motivation, is consistent with the cognitive revolution that has overtaken the behavioral sciences in the past several decades (Dyer, 1994).

1.2.6 Deception

The emphasis of conflicts of interest between individuals in the manipulation approach stimulated searches for deception in animal communication. Such searches dove-tailed with growing interest in the complexity of proximate mechanisms underlying communication. Smith (1965) had already noted flexibility in assessment mechanisms when he concluded that reactions to signals were based on both signal and contextual information. Is comparable flexibility evident in management mechanisms? For example, are signals invariably triggered by the same stimuli, or are there exceptions? A more flexible and perhaps more complex mechanism might be inferred if signals were 'withheld' in typical situations, or emitted in atypical circumstances, such as during deceptive activity. Marler, Karakashian and Gyger (1991) studied variation in food calling by male domestic chickens, in the presence of food but in varying social contexts. Food calling is used by males to attract females, often for courtship, and is typically associated with an offer of food to the female. Males varied their rate of calling depending on the audience present. Compared to the rate of calling with no audience, males elevated their calling rate in the presence of females, but reduced their calling rate when only another male was nearby. The authors noted that the inhibition of calling in the presence of male audiences raised the possibility that signals can be produced 'at will,' and therefore might be used deceptively.

Evidence for deception is not, however, necessarily evidence for flexibility or complexity in underlying mechanisms. For example, the form of deception found in interspecies mimicry could be founded upon fairly simple mechanisms. An alligator snapping turtle catches prey by luring it into its gaping mouth with its tongue, the tip of which mimics a worm. Worm-eating fish are attracted into the turtle's mouth by the twitching, two-pronged appendage covered with red spots (Wickler, 1968). It is unlikely that the snapping turtle developed its simulated worm with the 'intent' to mislead. (But the cognitive processes involved in deploying the 'worm' remains an open question.). In an effort to bring clarification to this literature, Mitchell (1986) proposed four categories of proximate mechanism, organized into a hierarchy of increasing psychological complexity. Deception could be effected by: (1) appearance, as in many visual mimicry systems; (2) a fixed response to a stimulus – producing a deceptive signal each time a particular stimulus occurs; (3) a modifiable response to a stimulus – using a signal deceptively because its effectiveness has reinforced such usage; and (4) planning – deploying a signal with the intent to deceive. The snapping turtle system would probably fit into categories one or two. Marler *et al.* (1991), on the other hand, appeared to be proposing that the food-calling system of chickens fits into category four.

1.2.7 Ontogeny

Tinbergen realized that the proximate mechanisms underlying communication change as individuals mature. These changes are revealed in age-related modifications in the ways that individuals participate in communication. The pattern of these changes differs for the different aspects of communicative behavior identified in a semiotic approach: (1) structuring signaling behavior (syntactics), (2) linking signaling acts to specific circumstances (semantics), and (3) using signaling as a tool for dealing with the behavior of others (pragmatics). Konrad Lorenz's views on development at the syntactic level have already been mentioned; he called the most basic structures of animal communication 'fixed action patterns' and felt that social learning had little to do with their ontogeny. With some important exceptions, the developmental part of his view still appears to apply to many vertebrate vocalizations. However, there is more to the syntactic level than just signal structure; combining different signals is also a syntactic phenomenon, and may be more dependent on learning. Moreover, little was said in the early literature about the role of social input in development at the semantic and pragmatic levels.

1.2.8 Exceptions to Lorenz's conclusions: the role of social companions in development of the structure of vocalizations (syntactic level)

The sounds uttered by birds offer in several respects the nearest analogy to language, for all members of the same species utter the same instinctive cries expressive of their emotions; and all the kinds that have the power of singing exert this power instinctively; but the actual song . . . (is) . . . learnt from their parents or foster-parents. These sounds, as Barrington has proved, 'are no more innate than language is in man.' The first attempts to sing 'may be compared to the imperfect endeavour in a child to babble.' The young males continue practising . . . for ten or eleven months.

(Darwin, 1871/1981, i , p. 55)

W. H. Thorpe (1954) and Peter Marler (1956; Marler & Tamura, 1962) initiated the modern study of birdsong development, and the role of learning. Marler's summary of song development in white-crowned sparrows, based in part on his laboratory studies, became the prototype for illustrating the interplay between internal and external factors in ontogeny (Marler & Tamura, 1964; Marler, 1970). Singing is primarily a male activity in this species. A young male spends the first few months of his life among singing males, but does not himself sing until about five months of age. His first efforts are not much like the final adult version, but his singing becomes progressively more adult-like over a period of about two months. Laboratory studies indicate that two kinds of experience are necessary for normal song development – hearing others sing normal song (sensory learning), and hearing oneself sing (motor learning). When tape-recorded song is used as the source of song tutoring (rather than live, singing males), the following patterns emerge. (1) These males do not acquire the songs of any other species (such as Lincoln's sparrows and song sparrows), but readily acquire white-crowned sparrow song, even when it is alternated with that of another species. (2) The sensory-learning phase is restricted to the period from 10 to 50 days of age. (3) Song makes the transition to normal form during the first singing period, from five to seven months of age, only if the singer can hear himself sing (Konishi, 1965). (4) This motor-learning phase is restricted to the first year; song structure crystalizes at this time, and shows little subsequent change.

The ethological concepts of imprinting and companions gave rise to a body of literature based on attachment theory. This provided a framework for understanding the development of many kinds of pragmatic competence in addition to companion recognition. Bowlby (1969) proposed that an imprinting-like process also underlies the development of human infants' attachment to their parents. As a result of early interactions with the parent (typically the mother), infants come to recognize their mother and feel most secure in her presence. The attachment system is a regulatory mechanism, Bowlby proposed; that is, it functions to regulate the infant's sense of security. This system has been shaped by natural selection so that the infant feels most secure in the presence of stimuli provided by the parent, because these stimuli have been associated evolutionarily with the many resources that the infant needs for normal development. Much of the infant's early communicative behavior, such as crying, babbling, and smiling, is the output of the attachment system which serves to maintain a sense of security by keeping the mother nearby. The infant learns how to regulate its security by developing more sophisticated ways to manage its mother's behavior, and comes to use the mother as a secure base from which to initiate exploration of the world. Ultimately, the characteristics of the mother as a caregiver are internalized by the developing individual, and become a working model of companions, upon which the individual's subsequent social relationships are founded. Attachment theory subsequently proved useful in understanding the development of social competence in nonhuman mammals (see Kraemer, 1992, and associated commentary for an overview). Normally-reared rhesus monkeys, for example, are much more competent than individuals reared without a mother in assessing the relative status of social partners, and using social alliances to achieve their own ends (Anderson & Mason, 1974).

1.2.11 *The salient points about attachment theory: self-interested social regulation, and age specificity in development*

The important points about attachment theory are that it identifies the relationship between parent and offspring (1) as the first context for communication by young of many taxa, and (2) as critical in the development of the ability to manage the behavior of others. (3) More broadly, attachment theory treats organisms as regulatory systems whose behavior serves to maintain conditions conducive to survival, and ultimately to reproductive success. It is argued in Chapter 2 that such a regulatory view provides a heuristic framework for the study of communication. (4) Finally, attachment theory offers a way of thinking about development that complements the typical approach in the animal communication literature (Galef, 1981; Owings & Loughry, 1985). Galef pointed out that the usual way to view development is as a process of accruing adult levels of proficiency: infants start out incompletely developed or incompetent, and become more completely formed as they mature. This might be called an adult-focused perspective. Galef used a parasite analogy to illustrate the complementary approach: infant mammals are more than incompletely formed adults, they are 'parasites,' i.e., consummate exploiters of the niche provided by the parent–offspring relationship. This might be called an 'age-specific' perspective, in that it proposes that infant and adult behavior differs, not just because the infant is incompletely formed, but also because the infant's behavior is appropriate for its current developmental stage. Youngsters cope with the world not in the more unilateral way of adults, but *through* the parent.

Attachment might seem to be all-encompassing; however, we acknowledge that it does not apply where there is little or no parental care, such as in many amphibians and reptiles. Nevertheless, the idea that commucation involves self-interested social regulation still applies.

2

The roles of assessment and management in communication

The input of the sensory nerve is not the basis of perception as we have been taught for centuries, but only half of it. It is only the basis for passive sense impressions. . . The active senses cannot simply be the initiator of *signals* in nerve fibers or *messages* to the brain; instead they are analogous to *tentacles* and *feelers* [last two italics added]. And the function of the brain when looped with its perceptual organ is not to decode signals, not to interpret messages, nor to accept images. These old analogies no longer apply. . . The perceptual systems, including the nerve centers at various levels up to the brain, are ways of seeking and extracting information about the environment from the flowing array of ambient energy.

(Gibson, 1966, p. 5)

Historical antecedents are discussed in Chapter 1. This chapter introduces what communication encompasses. Not all historical concepts and ideas described in Chapter 1 are included. The goal here is to provide a new synthesis to guide future research, emphasizing the central role that assessment and regulatory processes play in communication.

2.1 Placing assessment on a par with management

Many species of small temperate birds form winter feeding flocks, sometimes consisting of several species (Sullivan, 1984), which they maintain by repeatedly *contact* calling. The downy and hairy woodpeckers, black-capped chickadees, tufted titmice, and white-breasted nuthatches that forage together often emit such contact calls throughout the day, but may cease abruptly with signs of danger, such as when a flock member emits an antipredator call, or a hawk is detected.

Emitting contact calls probably benefits the caller by providing access to the advantages of flocking, such as reduced vulnerability to predators and enhanced foraging efficiency. Cessation of contact calling when danger is detected probably reduces the caller's conspicuousness to the source of danger. But, active assessment processes have given contact calling a significance beyond these presumed original individual benefits. Downy woodpeckers have come to capitalize on the negative association between contact calling and danger, treating the presence of such vocalizing as an 'all-clear' cue. Solitary foragers become less vigilant when they hear play-backs of contact calling by species which typically flock with downy woodpeckers, but do not reduce their vigilance when calling is presented that was recorded from species which are not usually their flock mates. Also, solitary foragers resume normal foraging more quickly in the aftermath of a hawk encounter when the calling of typical flock mates is played back than when it is not.

Chapter 1 described how Amotz Zahavi initiated interest in the significance of assessment in communication. Animals, as assessors, make adaptive behavioral adjustments by being selective, directing their attention to the most reliable cues available for appraising individuals and situations. These cues may or may not be based on signals, and even when they are founded on signals, they may be putting the signal to some novel use, as the downy woodpeckers above have. In spite of contributions like Zahavi's, the role of assessment is still underplayed in the study of communication. Therefore, terminology is adopted here that gives assessment a more explicitly active role. So, rather than using the passive terms receiver or reactor (e.g., as in Smith, 1977; Dawkins & Krebs, 1978; Gustafson & Green, 1991), the term *assessor* is used in order to highlight the active importance of this role. A major theme of this book is that recognition of the true importance of assessment will generate a fundamental change in how we think about communication. The term *manager* is used to refer to the sender, previously assumed to be the primary active, controling role in communication.

Even when considering assessment, rather than management, an assessment/management (A/M) approach leads us to begin our inquiry with the regulatory problems that the individual currently faces. We start with the regulatory problem because the bottom line in assessment is pragmatic, just as it is in management. Regulatory/managerial problems focus assessment and comprise its context. Pragmatically, assessment involves adjusting behavior to fit current circumstances. Equally pragmatically, management is the adjustment of behavior to bring current circumstances into line with the manager's interests (Hennessy *et al.*, 1981).

2.1.1 The implications of active assessment for communication

What does *active assessment* mean? The influence of assessment is pervasive. As Gibson (1966) implied, individuals live in a tangle of 'feelers'
maintained by others as channels for their assessment activities. Virtually
any activity is likely to 'catch' in this tangle, incidentally 'tugging' on
some feelers and thereby generating consequences for the individual.
The important consequences are likely to be harmful *or* beneficial and
their balance will affect communication.

Assessment forms the foundation upon which communication systems
are built. Assessment may be based on anything that another individual
does, e.g., eating, breeding, excreting, fleeing, flying, fighting, sleeping,
grooming, walking, etc. If it is used regularly for assessment, it becomes a
potential communicative activity such as a signal. A signal is unlikely to
evolve if a balance in favor of benefits is not available at the outset to
both assessment and management. Because assessment is so important,
we describe a signal *as an act deployed to capitalize on assessment systems.*
Communication results when the interplay between assessment and management has reached some degree of momentary predictability or stability in the balance between costs and benefits to participants. With most
vocalizations, we are beyond these initial steps. Most vocalizations are
already signals because they are produced by specialized vocal organs
highly modified for their functions. However, the same conditions are
prerequisites for retention or modification of signals. The effects of
assessment may modify management in the manager's lifetime or, ultimately, through natural selection. Management is an inevitable consequence of assessment, and communication systems are one result.

The founding of management on assessment systems can be illustrated
with Batesian mimicry, a form of between-species communication – but
the logic applies equally well within-species; (see Ryan, 1994). In Batesian
mimicry, individuals of a harmless species, called the mimic, manage
individuals of another species, called the dupe, by exploiting the dupe's
assessment of a dangerous third species, called the model (inappropri-
ately, as becomes evident below). Dupes are usually predators, but may
even be pollinating bees duped by orchid flowers into responding to the
flowers as mates! The rattlesnake/burrowing owl/dupe system involves
hearing (Rowe, Coss & Owings, 1986). Rattlesnakes are the dangerous
models; they are venomous, and use their venom to subdue prey and to
defend themselves against enemies. Rattlesnakes use threat as an early
line of defense, advertizing their dangerousness by shaking specialized

rattles at the tip of their tail. Individuals with a history of contact with rattlesnakes often retreat when they hear this rattling sound. Burrowing owls are the mimic species that capitalize on the potential victims' auditory assessment systems. Burrowing owls defend themselves from enemies in part by producing a vocalization that is startlingly similar to the rattling sound of rattlesnakes (Fig. 2.1). The similarity is so striking that California ground squirrels from a population familiar with rattlesnakes treat the model and mimic sounds with comparable caution (Fig. 2.2).

There has been confusion over the source of natural selection – how does the mimic come to resemble the model? Mimicry is assumed to arise as predators weed out imperfect mimics through natural selection. During this process, the model's warning signal is assumed to be the standard against which the mimetic signal is compared. However, note that feedback from the dupe is actually the only source that the mimic species ever 'sees' regarding the structure of the model's aposematic

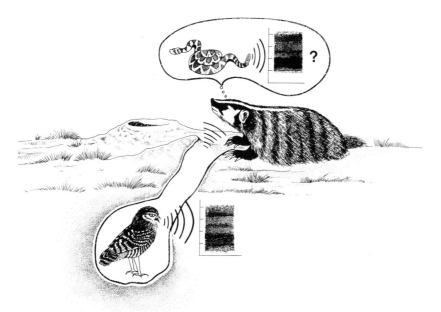

Fig. 2.1. A badger, about to seek prey by excavating a burrow system, has been halted by the resident burrowing owl's hiss. The owl's vocalization is remarkably similar to the rattling sound of venomous rattlesnakes, whose bite could injure the badger. As the spectrograms illustrate, both sounds are broad band, with regions of heavier emphasis at comparable portions of the frequency spectrum. When a relatively harmless species (burrowing owls) defends itself by copying a genuinely dangerous species (rattlesnakes), this is called Batesian mimicry. (Spectrograms courtesy of Matt Rowe.)

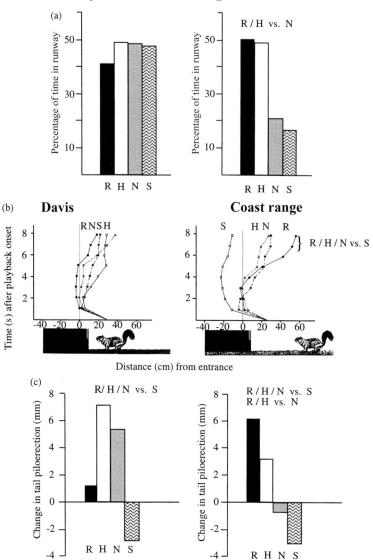

Fig. 2.2. A test of the hypothesis that burrowing owls are acoustic Batesian mimics of rattlesnakes. The responses of California ground squirrels from two populations were tested to four different sounds, each lasting 8 seconds and played back from a simulated burrow: rattling (R), hiss of owl (H), white noise (N), and scream chatter of owl (S). Squirrels from Davis are free of rattlesnake predation, and do not perceive rattlesnakes as especially dangerous. Those from the Coast Range experience rattlesnake predation, and treat rattlesnakes as dangerous. By two of the three measures illustrated, the rattlesnake-sophisticated Coast Range squirrels treated the rattling sound and burrowing owl hiss as

signal (see also Wickler, 1968; Towers, 1987; Guilford & Dawkins, 1993). Thus, it is the dupe's assessment system that shapes the mimetic signal most directly, not the model's signal.

The above point applies to all signals, not just mimetic ones. In general, it is biased responsiveness of assessment systems that sets the stage for some cue to be deployed in management. In most cases, the biased responsiveness arises from the 'discovery' of a cue that is useful in assessment.

Consider the complications that this could present for discovering the source of the relationship between a signal's structure and 'referent' (i.e., what assessing individuals infer from the signal; in this example, a rattlesnake). The case of burrowing owls mimicking rattlesnakes is straightforward, because the similarity between the owl's vocalization and the rattling sound of rattlesnakes is so obvious to us. However, in many other cases in which signals resemble some feature of the 'referent/model,' we cannot detect that resemblance, because our perceptual systems are not sensitive to the features that the target uses as cues to detect the referent/ model. Second, the signal–referent link might arise from referent-imposed constraints, such as the best way to deal with the referent. In this case, the cue may mimic other reactions to the referent (e.g., in being cryptic to avoid detection by a raptor), rather than being physically similar to the referent itself. It would be the cue's association with these and other activities, which occur primarily in the context of the referent, that would generate the signal–referent relationship. For example, the act of dropping to an inconspicuous low posture might become a visual signal for warning others about the presence of a raptorial predator.

This means that it is difficult to conclude with any confidence that the structure of *any* particular signal is arbitrary, because of the variety of sources of correlation between the presence of a cue and a predator. Therefore, conclusions that signals have arbitrary structure (e.g., as in

Fig. 2.2 (*cont.*)
more similar to each other, and more dangerous, than the other two sounds, whereas Davis squirrels did not categorize the sounds in the same consistent way. All significant planned comparisons are indicated above the bar graphs. Note that the other two sounds included white noise, which is as sibilant as the rattle and hiss. (a) Coast Range squirrels spent more time waiting in the runway leading to the sound-emitting burrow in response to the rattle and hiss than to the other sounds. Davis squirrels did not. (b) Coast Range squirrels stayed farther from the burrow opening in response to the three sibilant sounds than to the scream chatter. Davis squirrels did not. (c) Coast Range squirrels fluffed their tails more to the rattle and hiss than to the other two sounds. Davis squirrels did not. (Drawings courtesy of Matt Rowe.)

Cheney & Seyfarth, 1990) must be taken with a grain of salt, or even a whole shaker.

Placing assessment on a par with management involves exploring the limitations as well as the strengths of the two processes. For example, some signaling activities may be so physically demanding that physically limited individuals may have difficulty producing them effectively. As a result, such individuals may have difficulty 'faking,' for example, the possession of greater fighting abilities than they actually possess (Zahavi, 1977). The costliness of these physically demanding signaling activities may in turn be due to the quest for reliable cues for assessment, which would select against cueing on 'cheap' signals because they could be used even by individuals who could not follow through on a threat. In other words, the assessment process handicaps management by limiting the evolution of bluffable signals. However, costs are not confined to the signaling side of communication. For example, Dawkins and Guilford (1991) noted that '. . . so many displays used for quality advertisement involve duration . . . that it may be that receivers almost always have to pay time costs if they want an honest assessment.' Where the costs of complete assessment are high, then '. . . it will pay receivers to settle for cheaper, but less reliable indicators of quality instead.' (p. 865)

2.1.2 *Implications of active assessment for the role of semantic information in communication*

Scientists often use shorthand to simplify discussion of what are, in reality, complex and interactive processes (e.g., 'systems,' 'natural selection favors,' etc.). Usually, people are aware of constraints on the literal interpretation of the shorthand term. However, 'information' broke away from the constraints usually applied to shorthand when it was applied to communication after 'information theory' was developed in the late 1940s by communications engineers. Indeed, information became the most central term in definitions of communication (see Box 1.2). Here, it is shown where and how semantic information might be used properly. (Remember, from Box 1.2, that semantic information refers to knowledge, in contrast to the information of engineers, which translates to 'reduction of uncertainty.')

The proper place of semantic information is clarified through reference to the sharp distinction between management and assessment. Information is useful shorthand to describe what assessment produces. In strong contrast, management is focused on producing behavioral changes, not information. A shorthand term for the activity of assessing is the 'extraction of information.' When the active nature of assessment was not recognized explicitly, the transfer of information was needed to account for the impact of signals on the behavior of others. Information, sent by signalers and received by targets, took on an active, causal role. But, since individuals are already active in assessment, then nothing needs to be 'sent' by managers to make assessing individuals behave: *they are already behaving in their own interests.* Therefore, managing individuals try to influence *how* assessing individuals behave by affecting the conclusions they reach in assessment. Management is focused on pragmatic outcome, not information, both proximately and ultimately. And, because this process applies to many time scales, our 'active' management and assessment roles provide the same logical framework for both proximate and ultimate questions.

2.2 Management as regulation: applying the pragmatic, self-interested logic of natural selection in both proximate and ultimate time scales

What does it mean to apply the pragmatic, self-interested logic of natural selection to communication? As noted in Chapter 1, Burghardt (1970), Dawkins and Krebs (1978), and Krebs and Dawkins (1984) provided an answer for an ultimate time frame. Managerial activities are a product of their fitness consequences for the manager. Natural selection is the process of preserving those ways of managing others that yield the largest fitness benefits for the manager. Alfred Russel Wallace, the codiscoverer with Darwin of the principle of natural selection, drew an analogy between natural selection and a governor on a steam engine (Cronin, 1991). A governor regulates a steam engine's speed through a process of negative feedback. Similarly, natural selection limits deviations from the best-adapted state via differential reproductive success of individuals. Consistent with Wallace's suggestion, we propose that the proximate formulation most consistent with the logic of natural selection is regulation through feedback (Fig. 2.3a). Attachment theory (discussed in Chapter 1) provides such a formulation. It treats animals as negative-feedback systems that work to manage conditions in their own interests. As with natural selection, these regulatory processes, acting on individuals, produce both ultimate *and* proximate consequences (see also Burghardt, 1970).

A regulatory system can be thought of as a coordinated collection of parts whose activity tends to maintain some variable at a relatively constant value. Cannon's (1935) concept of homeostasis illustrates this idea;

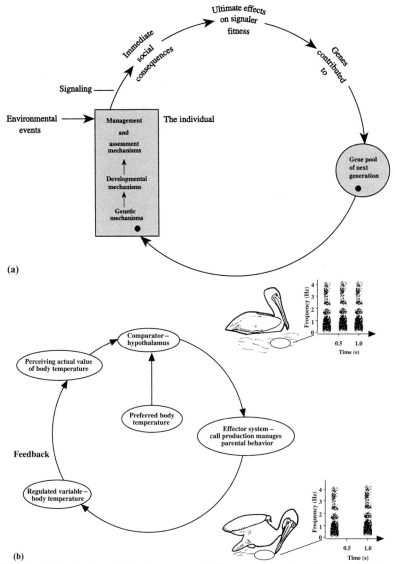

(a)

(b)

Fig. 2.3. Feedback in (a) ultimate, and (b) proximate time frames. Feedback processes are common in biology; they occur wherever circular loops of causation, such as those illustrated here, can be found. (a) This ultimate feedback is natural selection. An individual organism is an 'output' from the population gene pool. This individual emits signals to manage the behavior of others. The social consequences of these signals feed back on the population by influencing the individual's success in contributing to the gene pool of the next generation. (Redrawn from Alcock, 1989, with permission from Sinauer Associates.) (b) This proximate feedback is part of the embryonic white pelican's thermoregulation system. Embryos use squawk calls to elicit warming behavior by the parent, and slow their rate of squawking as their parent's response warms them. (Spectrograms courtesy of Roger Evans.)

he coined this term to refer to the mechanisms responsible for maintaining some condition of the body, such as its temperature, at a constant value. Some aspects of such regulation are based on processes internal to the body; internally generated metabolic heat, for example, can aid in raising body temperatures that have dropped below normal. Other aspects of such regulation depend upon use of environmental resources; an animal might cool off by moving to shade, for example, or warm up by moving into sunlight. Attachment theory has highlighted the idea that youngsters regulate their state by using parents and other conspecifics as resources. In other words, infants might regulate their body temperatures, for example, by regulating the behavior of their parents.

Regulation of egg temperature by birds involves negative feedback. Incubating parents monitor egg temperature, and take corrective action when the temperature deviates from optimal. However, as the time of hatching approaches and passes, parents monitor offspring less closely and regulation of their offsprings' temperature becomes less exact (Evans, 1990a; 1990b). Embryonic and neonatal birds begin to take a more active role at this time (Fig. 2.3b). For example, white pelicans of this age begin to emit squawk calls, vocalizing at higher rates as their temperature deviates further from about 37°C, and lower rates as the 37°C body temperature is restored (Evans, 1992). When the parents hear their young squawking, or even playbacks of squawking, they engage in brooding activities over their young which should result in adjustments in chick or egg temperature. Embryos are feedback sensitive; they can regulate their own temperature by calling when placed in an experimental incubator with a voice-activated heating and cooling system. Before hatching, squawking is the youngster's primary means of restoring preferred body temperature. After hatching, the young pelican adds shivering to its repertoire of means of producing heat to regulate body temperature (Evans, 1994).

Many regulatory systems appear to have internally specified 'preferred' values of the regulated variable, and act to minimize the difference between preferred and actual states of that variable. Many systems regulating physiological variables in animals, such as the one described above for egg temperature, are thought to work in this way (Toates, 1980). However, a specified preferred value is not a necessary feature of regulatory systems. For example, it is incorrect to think that evolution has a purpose, goal, or endpoint. Higher levels of organization or complexity may occur over time but these are not due to any special direction-giving process. No internally specified preferred value is involved in frequency-dependent natural selection, even though evolutionary stability results (see Chapter 1, and Maynard Smith, 1982). When a hawk strategy competes against a dove strategy, a commonly expected outcome is stabilization of the population at a constant mix of hawk and dove. Any deviation from these proportions increases the advantages of the reduced strategy and reduces those of the increased strategy, which reinstates the evolutionarily stable proportions.

In proximate regulatory processes as well, stability often results from interaction between system components rather than a set point. A very simple example of such a process involves two adversaries pushing against each other with equal force, resulting in the dynamic maintenance of a static position (e.g., Toates, 1980). Similarly, Geist (1974) has argued that some cases of noninjurious fighting may just appear to be based on individual decisions to fight conventionally. Both individuals may be attempting to injure the other, but each may parry the others efforts well enough to produce a proximate stalemate between offensive and defensive tactics. In these and many other cases, regulatory systems are actually higher-order entities, e.g., involving pairs of individuals, and the system's activity results from the interaction between two individual systems with their own, conflicting preferred values. The general message here is that points of stability in what happens between animals or strategies are not necessarily specified by the participating individuals; participants may be regulating other variables, or specifying different preferred values of the same variable, with stability emerging as a 'stalemate.'

2.3 The structure and functioning of assessment and management systems

At the 1973 International Ethological Congress in Washington, D.C., K. Nelson gave a memorable paper entitled: 'Is bird song music? Well, then, is it language? Well, then, what is it?' At least as plausible as either language or music is the possibility that bird song should be regarded as akin to hypnotic persuasion. . . But it may be that these are not all that different from each other. There may be a continuum between hypnosis as it is commonly understood and ordinary verbal persuasion, with the 'spellbinding' oratory of a Hitler or a Billy Graham falling between. There may be little difference between regarding bird song as music and regarding it as hypnosis. 'Hypnotic' rhythm and 'haunting' melody are clichés in the description of human music. The drug-like effect of the nightingale's song on the poet's nervous system ('a drowsy numbness pains my sens, as though of hemlock I had drunk') might be at least as influential on the nervous system of another nightingale.

(Dawkins & Krebs, 1978, p. 307).

2.3.1 The components of assessment systems

It is noted in Chapter 1 that Dawkins and Krebs, in the paper from which the above quote is taken, were contesting the utility of the idea that signals influence the behavior of others through information exchange. They were noting that traditional emphases have been on the effects of motivational states on signal production. In the language of information exchange, this would involve identifying the information that signals make available about the signaler's motivational state. Here, Dawkins and Krebs raise the complementary issue of signals as causes, of the motivational states of targets, rather than effects, of the motivational states of signalers. Their point is that animal signals may work like music does in humans, to influence the motivational states of listeners.

From an assessment/management perspective, we would say that Dawkins and Krebs had begun to explore the variety of assessment mechanisms in targets that management makes use of, to influence target behavior. They focused on motivational systems rather than information-processing, i.e., cognitive, systems. Motivational and cognitive systems are not mutually exclusive, nor do they cover the gamut of assessment mechanisms. In addition to cognitive and motivational systems, there are also perceptual and emotional mechanisms. The need for such diverse categories is indicated by the variety of mechanisms that underlie communicative behavior. These are all behavioral categories whose properties can be studied at the behavioral level, as well as at the level of their physiological underpinnings.

Perceptual systems include the sensory mechanisms with which we are all familiar, such as vision, olfaction, touch, and audition. These mechanisms support the actual pickup of cues needed for adaptive behavior. In studies of vocal communication, the sense of hearing is central. For example, female green treefrogs, *Hyla cinerea,* rely heavily on the sound of vocalizations of males as a basis for mate choice, as many frog species do (e.g., the Túngara frogs in the Prologue). Female mate choice is cued in part on the dominant spectral frequencies comprising the males' calls, and their behavioral preferences match those frequencies that are most stimulating to their auditory systems (Gerhardt, 1987). Indeed, female response to variation in several sound parameters indicates that they are most likely to approach those sounds that provide the greatest amount of stimulation to their auditory systems. Other examples from frogs are presented in Chapter 3.

with situations relevant to particular life concerns (Tooby & Cosmides, 1990; Lazarus, 1991) When work on a concern is facilitated, positive emotions are activated; when efforts are thwarted, negative emotions result. Emotions are the complex executive systems that orchestrate effective responses to such situations by coordinating the activities of perceptual, cognitive, motivational, and motor systems, focusing them on dealing effectively with adaptive problems.

Positive emotional arousal can be illustrated by the rewarding effects to a Norway rat of ingesting highly palatable food. This produces the positive state that Kent Berridge (1996) has called 'liking,' which is linked to the food's palatability and the sensory pleasure of eating. Liking has a neural substrate that is separate from that of the other aspect of reward, called 'wanting,' which consists of appetite or craving, and motivates the search efforts to gain access to more. A variety of activities falls under the heading of 'search,' including social assessment in which the further availability of food is discovered by attending to the food-associated calls of conspecifics (Marler, Dufty & Pickert, 1986; Hauser & Marler, 1993a; Caine, Addington & Windfelder, 1995), or by smelling the residue on other group members who have already located the source (Galef & Wigmore, 1983).

Negative emotional arousal can be illustrated by the strong reaction of adult humans to expressions of distress by human infants. For example, many readers are likely to have had the experience of having their enjoyment of a restaurant meal or theater film marred by the crying of an infant. Studies of adult responses to infant crying have been hard-pressed to identify any applicable positive adjectives for describing these sounds. The descriptors found to be relevant for rating crying sounds include how urgent, sick, arousing, grating, discomforting, piercing and aversive the sounds are (e.g., Gustafson & Green, 1989). Indeed, crying seems to manage adult behavior by simulating infant respiratory distress (Thompson, Olson & Dessureau, 1996), thereby motivating persistent efforts by adults to discover and alleviate the source of the infant's discomfort. These emotional effects are not confined to adult perceivers; crying itself is contagious, tending to generate the spread of crying even among newly-born infants in hospital nurseries (Hoffman, 1978).

Aversive infant distress vocalizations are not exclusively a human phenomenon. In many nonhuman primate species, infants have a call analogous to human crying that elicits maternal responses (Coe, 1990). These calls also appear to manage the behavior of adults in part by inducing a negative emotional state. For example, captive adult female squirrel monkeys were exposed either to a distressed infant or to the sight of the separation of a mother and infant. The levels of plasma cortisol in circulation were measured, as an index of arousal. (Cortisol is a hormone of the adrenal cortex, and a common measure of the stress response.) Both treatments induced significant elevation of circulating cortisol, relative to control stimulus conditions. (Similarly treated adult males also exhibited behavioral agitation, but no adrenal activation unless the infant was actually observed being held by a human.)

Considering the behavioral roles of perception, cognition, motivation, and emotion highlights the interplay between proximate and ultimate factors in the production of adaptive behavior. Natural selection favors those developmental processes that generate the ability to pursue the most appropriate course of action. But the production of adaptive behavior relies heavily on an individual's ability to assess its own proximate state and circumstances. This is true for several reasons. First, the timing and details of events in an individual's life are never entirely predictable, so the scheduling of activities cannot be preprogrammed. Second, since the interests of individuals conflict, the social cues needed for decision-making may be masked or suppressed, and thus require active extraction rather than just passive receipt. Third, many adaptive activities conflict with other activities. Foraging, for example, has the potential to interfere with antipredator behavior, in part because the allocation of attention to seeking food reduces its availability for scanning for predators. So, behaving appropriately often involves a set of compromises among conflicting demands, with the mix and urgency of demands varying constantly, requiring frequent assessment and adjustment of effort allocation. As Emlen *et al.* put it, 'It is now recognized that natural selection can operate on the *decision-making process* itself.' (Emlen, Wrege & Demong, 1995).

It is the interplay between management and assessment processes, in evolutionary, developmental, and immediate time frames, that generates the form of communication as we observe it. Whether the communication is founded on a balanced equilibrium between management and assessment, or is highly unstable depends on many complex variables. These include, for an individual, its age, sex, kinship relations with interactants, spacing, food, and other ecological requirements for life (and their distribution in space and time). Also for the individual, but not under direct proximate control, are its age-specific chances for future survival, demographic aspects of its population, competition with other species, probability of being eaten by a predator, what other conspecifics are doing and their relation to the sex and age of the individual under consideration, and a great deal of stochastic or chance elements like weather extremes or human activities. How communication is shaped by natural selection, ontogeny, and immediate processes – as a balanced outcome of manager and assessor self-interest (how the costs and benefits to both are balanced) – is a central organizing question for researchers to pursue.

2.3.2 Some of the ways that different components of assessment differentially influence communication

Even though the different components of assessment overlap and interact, distinguishing among the four is important. Because of their distinctive characteristics, they not only contribute differently to assessment, but also differ in their impact on management, via at least two routes. *Between individuals*: since managerial activities exploit the features of assessment systems, the four impose different kinds of constraints and opportunities on management. In other words, the features of managerial activities that best exploit cognitive systems, for example, may be different from the features of managerial activities that most effectively work through motivational systems. *Within individuals*: animals assess to uncover the cues they need for management, they do not seek knowledge for knowledge's sake. Therefore, the four components of assessment also constitute distinguishable types of influence on management by acting as different aspects of the mechanisms of management. Motivational systems, for example, judge how important an event is and therefore whether it should be dealt with. If the event is to be addressed, emotional systems evaluate whether it is positive, and should therefore be approached, or negative, and should therefore be avoided. Perceptual and cognitive systems work together to assess what precisely should be done about the event (e.g., it's a female – court! or, it's a male – threaten!)

Recognition of the roles of perception, cognition, emotion, and motivation in both management and assessment raises questions that have not routinely been addressed in the past. For example, most work on the role of emotion in communication follows Darwin's (1872/1965) emphasis by discussing the contributions of emotion to signal *emission* (management), (e.g., see Scherer, 1992). The formulation presented here also asks how emotion influences assessment, and how such effects can feed back on management (Klinnert *et al.*, 1983; Scherer, 1992; Owings, 1994). This issue is developed in more detail in Chapter 4; the following discussion is intended simply to illustrate some of these points.

For this discussion, motivation and emotion will be combined under the heading 'conation' in order to contrast their effects with those of cognition (see Owings, 1994). Conation can be defined as 'the act or faculty of impelling or directing muscular or mental effort' (*Webster's New Universal Unabridged Dictionary*, 1983), and can be distinguished from 'cognition,' which we defined earlier as all aspects of information processing beyond perception. The basic concern here is how conation and cognition differ in their impact on communication.

Targeting the conative components of assessment systems, compared with cognitive components, has the potential to produce unusually powerful and long-lasting influences. The social spread of snake-evoked fear in rhesus monkeys is a good example (Mineka *et al.*, 1984; Mineka & Cook, 1988). Rhesus monkeys captured as adults in the wild respond to snakes with many signs of fear, including fear grimacing, making threat faces, vocalizing, lip smacking, erecting the pelage, suddenly retreating, cage shaking, averting the gaze, and staring. Youngsters born in captivity do not respond to snakes in a fearful way, but they appear to 'catch' the fear of adults, a form of emotional contagion also described in humans (see Hatfield, Cacioppo & Rapson, 1994, for a review). That is, the young monkeys' reactions closely match the adult's fearful behavior even when the snake is not in view. When both snake and fearful adult are in view, captive-born young rhesus monkeys very quickly acquire an intense and persistent fear of snakes. This process may be similar to 'social referencing,' as described in human infants (Klinnert *et al.*, 1983), and 'cultural transmission' of enemy recognition in birds (Curio, Ernst & Vieth, 1978). In both of these additional cases, vocalizations play a significant role in the induction process.

What are the potential means and costs of inducing conative arousal in others? The above descriptions indicate that one effective method is for the manager to express that state also. That is consistent with the costs and benefits of management in many cases. An individual alarmed by a predator might very well benefit from inducing the same cautious behavior in relatives, or others whose welfare is beneficial to it. However, there are limits to the versatility of this managerial tactic, which may drive up its costs. When the managing individual is not experiencing the state, but needs to induce it in others, the state might be simulated by engaging in some of the activities typically associated with it. But, partial duplication of the behavioral syndrome is not as effective as complete duplication (see Vieth, Curio & Ernst, 1980), and may be difficult to do (Marler, 1984). The difficulty with emotional simulation arises from the many connexions among the components of emotional arousal, which results in stimulation of the rest of the conative system when one component is activated. For example, when humans read aloud a passage expressing a particular emotion, they induce more of that emotional experience in themselves when they also express that emotion in their vocal pitch and intonation patterns than when they read in an emotionally 'flat' way (Hatfield *et al.*, 1994). Such changes in the emotional state of the managing individual could impose additional costs of signaling by shifting priorities away from other important activities.

The vocal patterns of motherese (Fig. 2.4), demonstrated by Anna's mother in the Prologue, illustrate the utility of distinguishing cognitive and conative factors (Fernald, 1992). These distinctive melodies of speaking are the acoustic patterns that are most effective in regulating infant

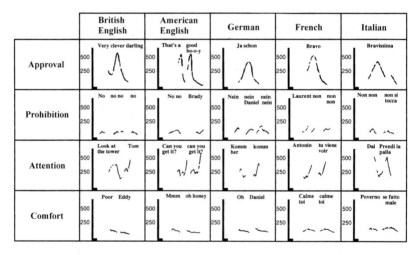

Fig. 2.4. Examples of the melodies of motherese, the distinctive patterns of speaking that adults of many cultures use to manage the behavior of infants. These patterns were illustrated by Anna's mother in the Prologue. Schematic spectrograms highlight the pattern of frequency modulation (FM) of the fundamental frequency of the parent's voice. Note the overall high pitch and wide range of frequencies covered in order to *approve* of the infant's behavior, or to get its *attention*. However, observe how the direction of FM at the ends of statements differs between these two managerial goals: declining for *approval*, but rising to get *attention*. The melodies of *prohibition* and *comforting* are strikingly different from *approving* and getting *attention*, involving low pitch and minimal FM. But, *prohibiting* and *comforting* also differ from each other. *Prohibiting* involves loud, abruptly onsetting sounds, whereas *comforting* sounds are soft and gradual in their amplitude changes. (Permission to copy from Fernald (1992) granted by Oxford University Press.)

behavior. Indeed, Fernald has argued that these patterns have been shaped, both phylogenetically and ontogenetically, by the motivational and emotional systems of infants, and reflect an adaptive adult strategy to capitalize on infant motivational and emotional systems. Later in the infant's first year, these distinctive acoustic patterns do begin to play a more cognitive role by contributing to language development. At that time, the cognitive matter of *what* the mother says begins to become more important. Initially, though, it is *how* she speaks, and the emotional and motivational effects on the infant that are most significant. Animal handlers also use motherese-like acoustic patterns to manage their animals' behavior (McConnell, 1991). Short, rapidly repeated, broad band notes are used to stimulate motor activity, and longer, continuous narrow band notes to inhibit movement. Experimental studies indicate that these are the most effective patterns for the differing purposes. As for human targets, some acoustic features seem designed to exploit the conative systems of nonhumans more than their cognitive mechanisms.

Do the features of management targeted on conative systems achieve their effects in different ways from features targeted on cognitive systems? Since cognitive systems are in part information extractors, we might say that management exploits cognitive processes partly by channeling the information-extraction activities of targets. How might we describe the sources of conatively mediated signal effects? Scherer (1992) has identified a substantial list of means whereby the induction of emotional arousal helps to sway targets. These means are explored more thoroughly in Chapter 4. Two items from that list will serve to illustrate the point to be made here. They are: (1) 'reducing information processing capacity . . . (in part through the) strengthening of unconscious automatic processing mechanisms,' and (2) 'eliciting phylogenetically preprogrammed behavioral action patterns.' These ideas may assist us in understanding the phenomenon of motherese; it is the *melody* of speech that is a critical determinant of a mother's effectiveness in soothing, reprimanding, or encouraging her infant. Perhaps the *patterning* of managerial activities is one feature that is especially important in the conative realm because it activates automatic processing mechanisms and phylogenetically old action patterns.

What is the evidence for the conative impact of patterning? A variety of animals appears to use rhythmic patterns of input to others as a means of managing their behavior (Schleidt, 1973; Owings, 1994), a phenomenon that is discussed below as *tonic communication*. Rocking patterns of movement are especially effective in quieting distressed human infants

(Byrne & Horowitz, 1981), and in reducing the arousal of nonhuman primate infants (Mason, 1971). External rhythmic inputs can in general have powerful effects on behavior through the process of *entrainment* (Decoursey, 1961; Timberlake & Lucas, 1989). Entrainment has been identified as an important mediator of communication between human mothers and infants (Davis, 1982); infant affective arousal becomes more positive as the infant's behavior becomes more synchronized with its mother's activities (e.g., Bernieri, Reznick & Rosenthal, 1988). When an infant is entrained to its mother, it might also be said to resonate to its mother. Resonance has been proposed as a general way of thinking about how animals tune into the environmental inputs needed for adaptive behavior (e.g., Michaels & Carello, 1981).

Application of the concepts of entrainment and resonance in a communicative context has typically involved a mutualistic view. In other words, individuals are thought to be striving to get synchronized as a means of harmonizing their endeavors (Bernieri *et al.*, 1988). While cooperation doubtlessly occurs, entrainment mechanisms may also provide conative 'handles' with quite a bit of leverage for self-interested management of the behavior of others. For example, the males of many species of insects, frogs, and toads attract mates by calling rhythmically (Greenfield, 1994), and in many of these the males call collectively with conspecifics, maintaining various patterns of coordination with the calls of each other. Adaptive explanations for most of these have focused on the benefits of combining the impact of multiple calls. However, coordination of calling is, for a subset of these species, a byproduct of individual males competing to jam each other's signals. Males reset the timing of their calling apparently to avoid closely following a nearby male's calls, briefly preceding that male's calls instead. The adaptive significance of such adjustments in the timing of calling lies in a bias in females' assessment systems called the 'precedence effect.' When two calls follow each other in close succession, females prefer the first of the two, even if its structure would otherwise make it less attractive. So, males adjust their calling to that of other males to give themselves the benefit of this precedence effect.

The discussion of rhythmicity and entrainment is just an example, of course, and one that falls closer to the motivational than the emotional aspect of conation. It illustrates the limitations of adopting a purely 'rational' approach to animal behavior, i.e., an approach centered on the idea of information. Neither rhythmicity nor entrainment relates in any obvious way even to such conative ideas as hierarchies of goals, and central life concerns, which capture the more cognitive aspects of conation (i.e., the ends toward which an animal strives). Of the many mechanisms of assessment that signals exploit, only a subset is describable in cognitive terms, of specified goals, etc.

2.3.3 The components of management systems

Since management systems are shaped in part by proximate and ultimate feedback from assessment systems, we should expect to find some correspondence between the components of the two types of systems. This should be especially true where our categories of components are useful ones in general for making sense of behavioral systems, as the categories are that we used for discussing the components of assessment systems, i.e., perception, cognition, motivation, and emotion.

Motivational frames for managing

An important key to understanding communicative behavior is to explore it as part of a broader motivational context. This is a major theme of this book. When we characterize what a signaling animal is 'doing,' we are often disposed to say that it is communicating. But, from the animal's perspective, it is often as appropriate to say that the individual is courting a mate, or seeking food, or maintaining its territory, or avoiding a predator, or regulating its body temperature, and is using signals as one means of facilitating that process. Focusing exclusively on communicative behavior is, from the animal's viewpoint, an artificial division.

The importance of considering broader motivational contexts can be illustrated with research on the causation of retrieval calling by preweanling rodents. Norway rat pups, for example, emit ultrasonic vocalizations when they become separated from their nest and littermates. ('Ultrasonic' sounds have a frequency above 20 kHz, the highest frequency audible to humans.) These calls elicit and guide retrieval by the mother, who returns her pup to the nest (e.g., Smotherman *et al.*, 1978). Although these have been called distress vocalizations, the critical causal factor in evoking such calling by the pup is the cooling that accompanies removal from the nest and littermates (Blumberg & Alberts, 1991). Indeed, the pup's ultrasonic sounds appear to have originated as a byproduct of heat-production mechanisms that function to counter this body cooling. Heat production is accomplished by an increase in the metabolic activity of tissue specialized for this purpose, called brown adipose tissue. Producing more heat requires more oxygen, which is made available through a distinctive respiratory mechanism known as laryngeal braking. In laryngeal braking, the larynx is constricted to maintain higher pressure in the lungs in order to increase oxygen transfer to the circulatory system. It is exhalation through the constricted larynx during breathing that produces the ultrasonic vocalizations.

So, the broader motivational frame for ultrasonic calling is thermoregulation. In fact, Blumberg and Alberts (1997) have argued that these sounds are not calls at all, in the sense of being part of a communicative system. These sounds, they propose, may be purely byproducts of laryngeal braking, and not specialized at all to induce maternal retrieval. Maternal assessment may carry the proximate burden of mother/pup interaction in this system, generating pup retrieval on the basis of incidental pup sounds. One way to deal with this question of whether these sounds have been favored by their communicative function is to seek adaptive relationships between the properties of these pup sounds and the adult auditory system that detects them. Such a relationship seems to exist: adults of several rodent species exhibit peaks in auditory sensitivity at those sound frequencies emphasized in the calls of their young (Brown, 1973). But this may reflect adaptation on the assessment side, not in management mechanisms. That is, the importance of ensuring pup survival may have selected for attunement of adult auditory systems to the properties of incidental sounds.

How might we test the above hypothesis that cold-induced pup vocalizations are not at all specialized for their effects on maternal retrieval? One way would be to explore in more detail the design features of the mechanisms involved in calling. We know that production and cessation are linked to heat production. But, there is also evidence that social factors modulate sound production, as they do food-call production in domestic fowl (see the discussion of tidbitting in Chapter 4). Testing pups in isolation from any cues arising from mother or littermates appears to activate separate physiological systems, one of which excites calling by the pup, the other of which inhibits calling. This dual effect has been hypothesized to be the product of the benefits (attracting mother) and costs (attracting predators), respectively, of calling (Hofer, Brunelli & Shair, 1994). The result is moderate elevation of calling in this context. However, calling is also contingent on social stimulation, which suggests design for communicative function. When contact with the mother is allowed even briefly in the testing chamber, and even when the mother is anesthetized, her presence inhibits calling by the pup, and her subsequent removal elevates calling well above moderate levels. This elevating effect occurs even when the mother has been cooled, so that the thermal effects of removing her are warming ones because her body is 12–14°C cooler than the testing chamber (Hofer, Brunelli & Shair, 1993). Such an effect may make adaptive sense. The exceptional potentiation of calling when there is evidence that the mother has been nearby may reflect the fact that the mother is more attractable under such conditions and therefore may arrive before any predators do who may also have been attracted by the pups' calls. So, the mechanisms underlying ultrasonic calling by isolated, cooled pups do appear to exhibit design features for communication.

Context

The term context has been used in at least two different ways in the communication literature (Leger, 1993). Those who have addressed the idea most systematically have done so from the perspective of the signal receiver (Smith, 1965; Leger, 1993). According to this view, signals are a source of information, but their meaning to perceivers depends on additional information available from other sources, such as events and conditions that precede or accompany the signal. Both Smith (1991a) and Leger (1993) confine their use of context to this meaning. The other way in which context is used comes closest to our treatment of the idea in this section. Context from this perspective is the situation in which the signal is emitted or its presumed function, e.g., courtship or agonistic vocalizations. This category of context is closely related to the broader motivational frame in which a signal is emitted, as discussed above. It refers to more than just the 'circumstances of signal emission.' It is the broader functional concerns of the signaler, the 'themes' that currently dominate its behavior. Context, in this view, is something that can be orchestrated by managers in self-interested ways.

Management and cognition

Remember from the discussion of the components of assessment that we use the term cognition broadly. Cognition refers to all input-processing activities beyond perception that support behavioral decision-making. We do not confine the term to processes that approximate human cognitive sophistication. So, we do not ask *whether* cognitive mechanisms underlie management; we assume cognitive processes and ask instead about the *level of functioning* of these mechanisms. In this sense, our approach is similar to Dennett's (1983), who considers multiple levels of intentionality in his discussion of the cognitive bases of animal behavior. Our regulatory analogy assumes intentions, that is, specifications

guish among the contestants, the ante might be raised to the next level in the aggressive sequence (a similar analogy is presented by Dawkins & Krebs, 1978). If an individual escalates immediately to the highest levels of threat rather than probing first, it may find out too late (i.e., after suffering a costly defeat or serious injury) that its rival is more dangerous than presumed.

According to this line of thinking, gradually escalating contests represent prudent assessment, not graded signals of the contestants' intentions. The emphasis on the information made available by signalers was an obstacle to discovering the insights provided by exploring the role of assessment in communication.

Units of management

Effective management of the behavior of others depends critically on the ability to vary managerial action. The more dimensions available for variation, the more precisely an individual can adjust its managerial efforts to maintain or move conditions toward preferred values. One way to vary managerial action is by exploiting different components of assessment systems, as discussed above. In addition, managerial action can be varied by adjusting the patterning of action at different temporal/ organizational levels. This can be illustrated by the following example.

Stonechats, songbirds of southern England, defend their dependent offspring when predators approach the nest, in part by emitting a long series of *whit* and *chack* calls (Greig-Smith, 1980). The ways in which they vary their *chacking* and *whitting* suggest that a complex regulatory process is involved. When a predator is detected near the nest, calling typically begins with a *whit*, and the most common subsequent pattern is to alternate *whits* and *chacks*. But, the call types are varied somewhat independently. For example, the rates of both *whitting* and *chacking* increase as a predator gets closer to the nest, but *whitting* declines as the predator moves away from nest, whereas *chacking* does not. *Whitting* also increases when a predator resumes moving after a pause, but *chacking* does not. *Chacking*, on the other hand, varies with the predator's movement direction, becoming more frequent when the predator moves away from or lateral to the direction of the nest, as compared to moving toward the nest.

More detailed investigation of the stonechats' antipredator behavior indicates why *whits* and *chacks* vary independently. The two calls are used to manage the behavior of different targets, specifically the nestlings and predator, respectively. The initial *whits* protect the young by suppressing their food-begging calls; food begging is loud and persistent, and may increase the youngs' conspicuousness to predators. Repeated *whitting* appears to be necessary because the young resume their food begging once the suppressive effects of *whitting* are removed. Maintaining the inhibition of food begging is an ongoing process, requiring periodic inputs to the young. *Chacks* appear to be targeted on the predator rather than the young, and are used to lure the predator away from the young. This is also an extended process that requires repeated *chacking*, as well as an intermittent increase in its tempo to 'reinforce' the searching predator's shifts to directions leading it away from the young.

Finally, the use of *whits* and *chacks* changes as the cycle of reproduction progresses. Prior to the eggs' hatching, the approach of a predator rarely elicits these vocalizations; but, both call types increase rapidly once the eggs hatch. Predator-evoked *whitting* and *chacking* reach a peak around the time that the young take their first flights from the nest, and decline rapidly thereafter.

The case of the stonechats helps us to understand how variation at multiple levels enriches the managerial potential of individuals. We highlighted a particular regulatory problem faced by parental stonechats; this might be characterized as the problem of keeping the distance between predators and offspring above some minimum value. But this problem actually entails two regulatory subproblems – managing the behavior of the predator, and of the young. Given the difference in the characteristics of both the two targets and the managerial goals for the two, it may not be surprising that they require different vocalizations. The broad spectral characteristics of *chacks*, for example, should make them easy to locate, a feature that should facilitate the predator-luring function of these calls. The narrower-band *whits*, on the other hand, are more strongly associated with perceived danger, and may therefore be more effective at suppressing the chicks' food-begging calls by inducing a fearful state in them.

Effective management by the stonechats in this predatory situation required adjusting more than simply which calls were emitted. In order to adjust to changes in the details of the predatory encounter, the stonechats also apparently needed to vary their rates of calling, separately for the two call types. Varying rates of *whitting* tracked the changing movements and location of the predator, whereas *chacking* tracked the predator's changing directions of movement.

Finally, the regulatory problems faced by adult stonechats changed as they progressed through their cycle of reproduction. Caring for eggs is different from caring for nestlings, because eggs do not make themselves conspicuous with loud, persistent food-begging calls. So, *whitting* in response to predators is unnecessary before the eggs hatch. Later, as the youngsters fledge and become more self-sufficient, they are less dependent on the parents' care. So, protective parental activities such as leading the predator away from the young become less useful, and *whitting* and *chacking* therefore decline.

In early views of animal communication, the capacity for adjustment in managerial output was thought to be quite limited, because most species have only a small number of displays available. Smith (1977) provided a review of the literature estimating repertoire sizes. While acknowledging the limitations of these approximations, especially the probability that repertoire sizes are underestimated, Smith nevertheless concluded that most species are unlikely to have more than 60–80 displays in their repertoires, and many are likely to have fewer (p. 172). This alone does not provide much room for the many adjustments needed to deal with the variety of situations the individuals of most species encounter. More recently, however, it has been recognized that switching to another display is only one of several means of adjusting managerial action (Morton, 1977; Owings & Hennessy, 1984; Smith, 1986b). Animals also have additional means of adjustment, including combining displays in various ways, as the stonechats do, and varying display structure. As discussed below, these additional means of adjustment greatly enrich an individual's managerial potential.

A structural taxonomy of dimensions of adjustment

The most tangible managerial units are individual vocalizations; these are readily viewed as coherent acoustic units in sound spectrograms. As noted in Chapter 1, the invention of the sound spectrograph made objective assessment of sound structure possible in the 1950s. This revolutionized the study of animal communication. However, it also focused attention on just one organizational level of vocal behavior, individual vocalizations, and so limited insights into the variety of ways that vocal behavior can be adjusted to changing situational demands. This point can be illustrated by research on the antipredator vocalizations of black-tailed prairie dogs. King (1955), for example, distinguished between the general warning bark, often used in the presence of terrestrial predators, and the hawk-warning bark. The general warning bark is a 'short, nasal yip that varies in intensity and frequency with the stimulus that produces it.' The hawk-warning bark differs in being 'faster, more intense, of higher pitch and of shorter duration.' King detected two levels of variation in vocal behavior – the structure of individual barks (e.g., more intense, of higher pitch), and the ways in which barks are combined (e.g., hawk-warning barks are faster). In an early spectrographic study of black-tailed prairie dog vocalizations, Waring (1970) detected no differences in the spectrograms of barks elicited by aerial and ground predators (but he offered no quantitative comparison). Rather than accepting combinations of calls as another legitimate level, Waring dismissed the brevity of the aerial-predator barks as a byproduct of the rapidity of the raptors' maneuvers. Such emphasis on the structure of individual calls, and inattention to other levels, has characterized most studies of vocal behavior.

Recently, however, recognition has spread that managerial activity is structured at multiple levels, loosely analogous to the ways in which human language involves phonemes, words, phrases, sentences, paragraphs, and so forth. For example, to make sense of the variety of forms of calling by the many species of ground squirrels, it is necessary to recognize, at a minimum, notes, series of notes, and series of series of notes (Owings & Hennessy, 1984). Similarly, Smith (1986b) has proposed a taxonomy of independent levels at which repertoires are organized; these include signal acts, variations of signal form, and patterned combinations of signal units.

Smith noted that signal acts represent the level we typically think of when the topic of animal communication comes up. These include the songs, chirps, and raised-crest displays of many birds, the arched back posture of defensive cats, the *barks* of dogs and *meows* of cats, *crowing* by chickens, and laughing and frowning by humans. The black-tailed prairie dog's repertoire of vocal acts includes *barks, rasps,* and *yips* (Fig. 2.5; Smith *et al.*, 1976; 1977)). These different calls are emitted in broadly different circumstances. *Barks*, for example, are used while avoiding or repulsing efforts by other prairie dogs to interact, and while responding to disturbances caused by predators such as hawks and coyotes, as well as by other heterospecifics (Fig. 2.6). *Yips*, on the other hand (Fig. 2.7), are associated with such situations as boundary challenges between territorial males, encounters with snakes (Owings & Owings, 1979), and the aftermath of disturbances from other prairie dogs or from raptors. Finally, *rasps* are used to threaten to attack another prairie dog, and occur most often in aggressive contexts such as when grappling, biting fights break up.

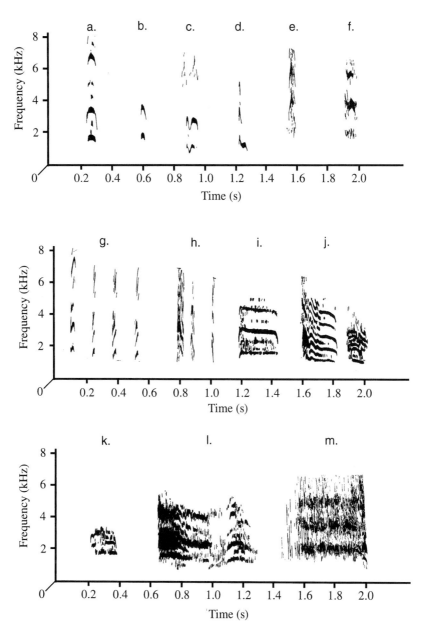

Fig. 2.5. Components of the vocal repertoire of black-tailed prairie dogs, illustrating variation both among and within types of vocalizations. (a–f) Variants of barks. (g–h) Chitter barks. (i–k) Yips. (l) Harsh yip, intergrading with (m) rasp. (Spectrograms courtesy of John Smith.)

Fig. 2.6. Black-tailed prairie dogs bark in a variety of situations, including encounters with mammalian predators such as the coyote illustrated here, as well as with avian predators and other prairie dogs.

Fig. 2.7. This *jump-yipping* black-tailed prairie dog is signaling by throwing its forequarters into the air, and then dropping back to a quadrupedal posture, while emitting a yip. *Jump-yips* may be emitted while dealing with a rattlesnake, as illustrated here, as well as at the end of a tense encounter with a hawk or another prairie dog.

It is clear from even this cursory description that each one of the calls of these prairie dogs is applied in a wide variety of circumstances. Does this mean that the structure of each call is imprecisely fitted to the situational demands of its use? Not necessarily. Very little perusal of the spectrograms of these calls is needed to discover quite a bit of structural variation within each category. *Barks*, for example, come in 'broad' (Fig. 2.5a) and 'narrow' (Fig. 2.5b), 'bi-peaked' (Fig. 2.5c) and single-peaked' (Fig. 2.5d) and 'clear' (Fig. 2.5a) and 'harsh' (Fig. 2.5e) forms. Similarly, *yips* come in 'harsh' (Fig. 2.5l) and 'clear' (Fig. 2.5j) forms, as well as 'single-note' (Fig. 2.5i) and 'two-note' (Fig. 2.5j) forms. Is it possible that this within-call variation in structure might reflect adjustments to the changing demands of the calling situation? Observations from several sources suggest that it may. The addition of harshness to calls, for example, seems to be associated with an increased likelihood that the caller will attack the target of the call, a pattern predicted by the motivation-structural rules hypothesis (Morton, 1977). Similarly, when *yips* are emitted while confronting a snake, their forms tend to be more bark-like when the snake is quite dangerous, compared to those cases in which the snake is less threatening to the yipping prairie dog (Owings & Loughry, 1985).

We cannot be certain of the communicative significance of such structural variation until we have explored its details, causes, and consequences. Consider, for example, the coo vocalizations of Japanese macaque monkeys (Fig. 2.8). Some of the variation in coos could reflect individual differences in the structure of that call rather than adjustments by an individual to its circumstances. This is because some of the listed situations are defined in part by the age or sex of the calling individual (e.g., young alone, female at young, separated male). Rhesus macaques, a closely related species, do not adjust the structure of their coo calls systematically with social context, but individuals differ substantially in the structure of their coos, cues that other assessing monkeys can use (Hauser, 1991; Rendall, Rodman & Emond, 1996). Individuals of some species, however, have been demonstrated to vary calls with social context, and this variation has been found to make a difference. Vervet monkeys, for example, use grunt vocalizations in at least four different contexts – when approaching a dominant group member, when approaching a subordinate, when observing a group member move into the open, and upon seeing another group (Cheney & Seyfarth, 1990). Playbacks of these variants of grunts evoke different responses, indicating that they differ structurally and make a difference in communication.

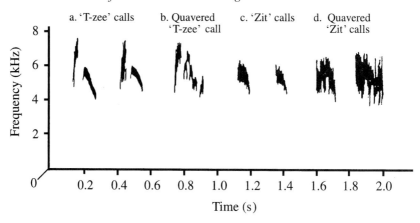

Fig. 2.9. Two levels of variation in the structure of the vocalizations of eastern kingbirds. (1) Switching between call types *t-zee* (a and b) and *zit* (c and d). (2) *Quavered* (b and d), and *unquavered* (a and c) forms of each call. (Spectrograms courtesy of W. John Smith.)

further in Chapter 3) provide a way to deal explicitly with the pervasive variation evident in the form of vocalizations, without sacrificing the very useful idea of repertoires of relatively discrete types of signals. The empirical utility of the hypothesis that this second level is independent of the first awaits further research. It is not clear how widespread independent repertoires are for varying signal structure. Nevertheless, the combination of M–S rules and Smith's subsequent formulation looks very useful right now.

Managerial activity can also be adjusted by varying how signal acts are combined. This level includes simultaneous emission of different signals. Humans, for example, may adopt angry facial expressions as they verbally reprimand their misbehaving children. Dogs may display their canine teeth as they emit threatening growls. The prairie-dog *yips* mentioned above are usually combined with a distinctive body movement, in which the caller 'leaps' to an extended bipedal stance and then drops quickly to a lower posture. Level three also includes sequentially combining different signal acts into compound displays. For example, male Siberian titmice serenade their mates during egg-laying. Their serenades consist of sequences of at least five different categories of notes, all of which are also used alone in nonserenade circumstances (Hailman, Haftorn & Hailman, 1994).

A temporal taxonomy

Many features of the world that we live in are patterned in time frames of various lengths. For example, day and night alternate with a cycle length of 24 hours; the moon cycles through its phases in about a month, generating not only significant modifications in nocturnal illumination, but also the waxing and waning of the heights of the oceans' tidal changes; and the approximately 365 days that the earth requires to revolve around the sun generate the annual cycle of seasonal change. Long-lived animals often adjust to such significant environmental changes by modifying their own activities. Optimal ways of behaving during the warmth and brightness of daytime, for example, are often quite different from the best ways to behave during the cool and darkness of night. Indeed, most animals are adapted for particular phases of day–night cycles, being either nocturnal (night active), diurnal (day active), or crepuscular (twilight active).

Given the many time frames in which our physical environments vary, it should not be surprising that most organisms pattern their activities in many time frames too, in part as a way of adjusting to their fluctuating environmental conditions. Such temporal activity patterns, in turn, become additional sources of patterning in the environments of other organisms in the same vicinity. A nocturnal prey species, for example, can be a source of selection for nocturnality in predators. Similarly, if this prey is also most abundant at a particular time of the year, the predator might be selected to schedule the birth of its young to coincide with this abundance. Thus, many temporal patterns of behavior are adaptations to the environment's patterns in multiple time frames.

The managerial systems that regulate accomplishment of major life tasks, such as feeding and reproducing, often include nested components that operate in different time frames (Timberlake & Lucas, 1989). Ethologists have long recognized these different levels in the organization of behavior in the distinction between appetitive and consummatory behavior (Craig, 1918). 'An appetite . . ., so far as externally observable, is a state of agitation which continues so long as a certain stimulus, which may be called the appeted stimulus, is absent. When the appeted stimulus is at length received it stimulates a consummatory reaction, after which the appetitive behavior ceases and is succeeded by a state of relative rest.' (p. 91)

Figure 2.10 is founded on Timberlake and Lucas's behavior systems approach, a modern and more detailed treatment of the appetitive–consummatory distinction. This drawing illustrates the hierarchical structure

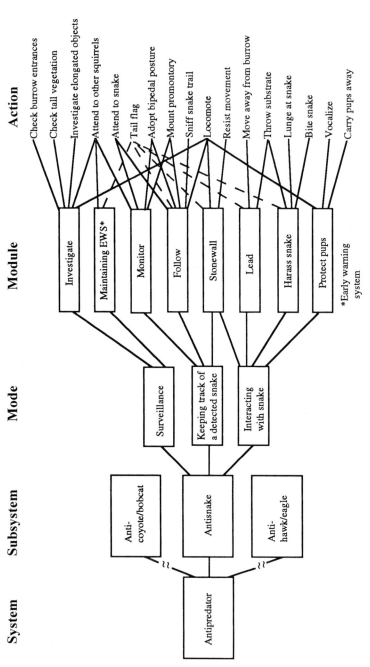

Fig. 2.10. An example of a behavior system's descriptive framework for dealing with the many organizational levels and time frames in which complex behavior is patterned. The example used here is the set of regulatory mechanisms in female California ground squirrels that function to deal with the threat of rattlesnake predation on her pups. This formulation is modeled after Timberlake and Lucas (1989). See text for additional details.

of the systems that regulate behavior, using the antirattlesnake behavior of California ground squirrels as an example. Since rattlesnakes eat pups rather than adult ground squirrels (Fitch, 1949), we treat this category of behavior as a subsystem of the parental care system of females. This subsystem illustrates behavioral patterning in broader time frames to the left (Modes – one level of appetitive behavior), and narrower time frames to the right (Modules and Actions – the latter representing consummatory behavior). Three *modes* are illustrated that differ in their proximity to an actual snake encounter. As pups are nursing and rattlesnakes become common, mothers move into a mode of *surveillance* for snakes (Hersek & Owings, 1993). If a mother detects a rattlesnake that is not close to her nursery burrow, she shifts into a mode of *monitoring* this snake that has not yet become a threat to her pups. If she is unskilful or unlucky, the snake will begin a search for her nursery burrow, which forces her into the mode of *coping* with an actual encounter with the snake (Hennessy & Owings, 1988). At the next level, consisting of narrower time frames, the mother deploys several behavioral *modules* in each of these modes. In the surveillance mode, for example, she *watches* directly for the snake and also seems to *maintain an early warning system* (Hersek & Owings, 1993). Similarly, while *coping* with an encounter, she may *stonewall* to obscure the nursery burrow location if the snake has not located it, or *harass* the snake if it has located the burrow (Hennessy & Owings, 1988). *Action* patterns are the smallest units, which therefore are completed in the shortest time frames. Each *module* deploys a number of different *action* patterns, many of which are shared by multiple *modules*. The signaling action called *tail flagging*, for example, is used both to *maintain an early warning system*, and to *lead* the snake away from the nursery burrow.

Participation in communication is a means whereby individuals maneuver their way through many of their life tasks. Indeed, a major theme of this book is that a key to understanding communication is to study it in the context of the broader regulatory systems that communicative activities serve. Therefore, we should find that communicative activity, like other biological activities, is patterned in multiple time frames. Tail flagging, for example, is deployed in a more extended time frame for the early warning system module, and a more immediate time frame for the snake-leading module. (See the discussion below of tonic and immediate time frames, respectively.) This difference is expected, since early warning is associated with the more prospective surveillance mode, whereas leading is a part of the mode of coping with the immediate presence of a snake.

How biological processes transpire at different 'rates' has already been discussed. This line of reasoning is extended below, where it is proposed that Tinbergen's four questions are best thought of primarily as involving different time frames. For the moment, we simply note that Tinbergen's two categories of proximate time frames, immediate causation and ontogeny, are insufficient to deal with the variety of frames evident in communication. Specifically, tonic communication is discussed to make the point that many communicative processes involve proximate time frames intermediate between Tinbergen's two. This should make it clear why a formulation involving just two proximate time frames, or even three, is insufficient for a complete understanding of communication.

Immediate time frames

Various regulatory problems demand an individual's attention, time, and energy. These problems differ from each other in many ways, including the level of urgency they typically pose for the individual. Motivational systems have been shaped by selection to ensure that an individual's efforts meet the distinctive properties of each problem (Simon, 1994). For example, individuals often treat the arrival of a predator as an urgent matter. Feeding, on the other hand, is often set aside temporarily, in order to deal with a predator, for example. The scheduling of feeding is more flexible because significant amounts of energy can be stored by the body; this makes the absence of food a less immediate threat than the presence of a predator. Perhaps these considerations help us to understand why so much empirical work on vocal communication has focused on 'alarm' calls, that is, calls that function to deal with the detection of a predator (e.g., see Macedonia & Evans, 1993). Since antipredator systems usually have high priority among motivational systems, such vocalizations are often readily evoked through experimental simulations of predatory encounters. For the same reasons of priority, the playback of such vocalizations often evokes immediate and distinctive reactions. So, alarm-call systems are among the easiest to study experimentally. And, the immediacy of their associated circumstances makes them prime examples of systems that emphasize immediate time frames.

The antipredator behavior of stonechats, described earlier, illustrates the patterning of vocal behavior in immediate time frames. Changes in the rate of calling constituted immediate adjustments to modifications in the details of the predatory encounter, including how close the predator was to the nest and whether its movements were taking it closer or farther from the nest. Most of the work on vocal communication has concentrated on such relatively immediate time frames, i.e., those adjustments that occur within seconds or minutes of a situational change.

Tonic time frames

Even predatory situations have longer-term dimensions. Bobcats, for example, are important predators of California ground squirrels; their arrival at a squirrel colony evokes immediate flights to refuge, as well as chatter vocalizations that cause other squirrels to bolt to the safety of their burrows (Owings & Leger, 1980; Owings *et al.*, 1986). But, bobcats also make use of ambushing as a predatory tactic, settling into hiding in nearby brush after arriving, and launching an attack once the squirrels have resumed their previous activities, such as foraging. This form of predatory behavior extends the time frame over which danger persists after the predator has ceased to be apparent. This is a source of selection and experience that should favor more tonic forms of antipredator behavior. One probable result has been the rhythmic or repetitive pattern of calling that squirrels use in the later parts of an encounter with a bobcat, as well as in its aftermath. This is an example of tonic communication. Tonic signaling involves adjustments in broader time frames, i.e., tens of minutes, hours or days, rather than seconds or minutes.

If the effects of repetitive calling are sought in an immediate time frame, these calls seem nonfunctional; it is rare to note a discrete response to an individual call in a repetitive series. A more appropriate measure of the impact of repetitive calling would be one involving sustained states, rather than discrete reactions. Such a measure uses a time frame that matches the temporal patterning of repetitive calling. Tonic signaling is founded upon the cumulative effects of repeated vocal inputs to targets (Schleidt, 1973). Indeed, squirrels listening to repetitive calling exhibit an elevated state of vigilance relative to control observations when no repetitive calling is audible (Loughry & McDonough, 1988).

The logic applied here predicts that a relationship should be detectable between the temporal persistence of threat posed by a type of predator, and the proportional emphasis of tonic and immediate time frames in signaling. Rattlesnakes pose the most temporally sustained threat to California ground squirrels – these predators may move into the vicinity of a particular burrow system and stay for days (Hersek, 1990). Perhaps because of this distinctive source of selection, these squirrels exhibit exceptional emphasis of tonic time frames to deal with these snakes. In a study of the signal used most often to deal with snakes – tail flagging – 90 percent of the incidents of this signal were used tonically, that is, outside the immediate presence of snakes, but in circumstances of elevated snake risk (Hersek & Owings, 1993).

Exploitation of tonic time frames may be the most common pattern of vocal communication. This pattern of communication may be underrepresented in the literature for a variety of reasons, including the subtlety of the signals and of their behavioral effects on targets. The units of tonic communication are also longer/larger; so, more time is required to collect a large sample size.

In closing this section, we must note that signals cannot always be separated unequivocally into tonic or phasic categories. Most signaling events have both immediate and longer-term effects on others. The grunts used by mountain gorillas provide an example (Stewart & Harcourt, 1994). These large primates live in small cohesive groups that alternate between periods of travel and feeding, and sessions of resting. Grunts are emitted throughout resting sessions, and about half of them have the immediate effect of eliciting an answering grunt from another group member. At the same time, they appear to have a more tonic effect of building a consensus that it is time to resume movement and foraging. Grunts are emitted only infrequently at the beginning of a resting session, and most frequently during the minutes preceding a group departure.

Developmental time frames

Participating in communication may have not only immediate and tonic consequences, but may also generate relatively permanent modifications in individuals. These latter adjustments are called developmental; they are very long-term proximate adjustments to changing internal and external conditions. European blackbirds, for example, 'mob' certain types of predators by perching in their vicinity and flying back and forth past them, while emitting *duk* calls and flicking the tail and wings (Vieth *et al.*, 1980). When blackbirds perceive other blackbirds behaving in this way toward a previously neutral object, the perceiving birds 'catch' this concern about the object, responding to it by mobbing during that and future encounters. In laboratory studies, the effects of perceiving the mobbing of an object are powerful enough to induce blackbirds to mob not only a stuffed nonpredatory bird (a male noisy friarbird), but also such innocuous objects as a plastic bottle. Playing back a recording of *duk* calls alone in association with the stuffed friarbird, with no accompanying visual components of mobbing, is sufficient to induce the blackbird to treat the friarbird as a predator.

The topic of development is discussed in much more detail in Chapter 4, where it is emphasized that not all developmental change is of the 'cumulative' sort discussed above, in which individuals learn progessively more about their environments. Some changes in communicative behavior are more like the metamorphoses observed when a frog tadpole makes the transition to adult form. Where changes are metamorphic, such as when a bird fledges or a mammal is weaned, the significance of particular signals can change radically. In such cases, communicative behavior is considered to be age specific, and makes sense only in light of the characteristics of that developmental stage.

2.4 Rejoining proximate and ultimate: implications for vocal communication

In Chapter 1 proximate and ultimate questions were contrasted, noting that they deal with causal effects within and across lifetimes, respectively. While this distinction seems simple to understand, this was not always the case (Mayr, 1982, pp. 72–3):

All biological processes have both a proximate cause and an evolutionary cause. Much confusion in the history of biology has resulted when authors concentrated exclusively either on proximate or on evolutionary causation. For instance, let us consider the question, 'What is the reason for sexual dimorphism?' T. H. Morgan castigated the evolutionists for speculating about this question when, as he said, the answer is so simple: male and female tissues during ontogeny respond to different hormonal influences. He never considered the evolutionary question why the hormonal systems of males and females are different. The role of sexual dimorphism in courtship and other behavioral and ecological contexts was of no interest to him.

Obviously, clarification of the proximate/ultimate distinction eliminated some counterproductive debates in biology, but the dichotomy eventually became a problem in its own right (Dewsbury, 1992). On the positive side, the study of behavior became more complete as research on the adaptive significance of behavior spread. Eventually, however, the proximate/ultimate distinction became so canonized that it developed into an obstacle to the synthesis of the two. They were said to be logically distinct, i.e., proximate questions were about *how* behavior worked, whereas ultimate questions dealt with *why* behavior was designed the way it was (e.g., Alcock, 1989). And, they were proposed to involve incompatible methods (e.g., Beer, 1982). Proximate and ultimate became separately addressable questions, rather than being pursued together.

The coordinated study of proximate and ultimate processes is facilitated by adopting a regulatory approach, as discussed above. A regulatory approach emphasizes the many features that are shared by the two classes of processes, along with the few that distinguish them. For example, proximate and ultimate questions are not logically distinct; a regulatory approach makes it clear that *how* and *why* questions can be answered both proximately and ultimately. We can ask *why* white pelican neonates which are too cold behave the way they do. The proximate answer is to increase their temperature, a proximate consequence to which the youngsters are immediately sensitive. We can also ask *how* they affect their temperature. A partial proximate answer is that they evoke parental brooding behavior by squawking. In an ultimate time scale, we could also ask *why* cold neonates behave the way they do. The answer would lie in the impact of squawking on the lifetime fitness of the individual. Answering the ultimate *how* question would involve determining the intricate cascade of events leading from squawking, through effective thermoregulation, to success in reproduction.

A second step toward synthesis of proximate and ultimate would involve expanding this dichotomy to Tinbergen's four questions. Remember from Chapter 1 that Tinbergen's four questions can be split into two types of proximate questions (causation and ontogeny), and two types of ultimate questions (survival value/function, and evolution). Both the proximate–ultimate distinction and Tinbergen's four questions are best considered as referring to different time frames in which regulatory processes operate, rather than to logically different types of processes. Thus, proximate questions are about processes having their effects within the lifetimes of individuals, with Tinbergen's question about causation reflecting shorter time frames and his question about ontogeny referring to longer frames, but both within lifetimes. Ultimate questions, on the other hand, are about across-lifetime effects, with Tinbergen's question about function referring to relatively immediate ultimate time frames (the current evolutionary utility of the trait), and his question about evolution referring to longer ultimate time frames (e.g., the evolutionary process of deriving the current trait from some older one).

How does this expansion facilitate synthesis? It highlights another distinction between types of questions, other than operative time frame, that is embedded in Tinbergen's taxonomy of questions. The two questions that Tinbergen addresses first are those about the mechanisms that are the immediate causes of behavior, and the sources of selection that account for the current utility of behavior. Neither of these is a fundamentally historical question; they deal with unchanging behavioral mechanisms and sources of selection. In contrast, questions about evolutionary history and development are fundamentally historical; they focus on how behavior changes through time. Considerations of history, whether developmental or evolutionary, direct our attention to how an animal's past structure is revealed in present design, and how current structure constrains and channels future changes.

Current structure involves both the mechanisms that guide behavioral development and those that regulate behavior. Developmental mechanisms function to regulate production of behavioral mechanisms appropriate for an animal's needs and circumstances. This means that evolutionary modifications in behavioral mechanisms must be accomplished through adjustments in developmental mechanisms (Gould, 1977; Mason, 1979b). So, the nature of the programs that regulate development need to be understood if we are to comprehend what the limits and opportunities are for evolutionary change in behavior.

This point can be illustrated with research into the role of developmental mechanisms in the evolution of sex differences in singing in the genus of wrens *Thryothorus*. This genus includes species such as the Carolina wren in which females do not sing at all, the rufous and white wren (*T. rufalbus*), in which females sing less than males, and others in which there are few sex differences in singing (the bay wren, *T. nigricapillus*, and buff-breasted wren, *T. leucotis*) (Arnold *et al.*, 1986). For the time being, the question of why such diversity might have evolved, i.e., what the sources of selection are that may have favored these modifications, will be set aside. The focus here is on how these evolutionary changes may have been mediated. As mentioned in Chapter 1, heritable phenotypic variation must exist for natural selection to work. How might that variation be generated? Arnold (1994) has proposed that an understanding of proximate processes, specifically song ontogeny, might help answer this question. For example, early exposure to the hormone estrogen initiates development of the complex set of neural structures that support singing. Perhaps, Arnold hypothesizes, the small genetic changes that produce congeneric species have led to different amounts of early estrogen secretion. If, for example, the primitive state were singing by both sexes, then selection would have already shaped developmental mechanisms that expose both sexes to high estrogen levels early in ontogeny, initiating development of neural mechanisms for singing in both. If selection subsequently changed, favoring singing only by males, this change could be mediated by selection for mutations that reduce early exposure of females to high estrogen levels. The result would be development of neural mechanisms for singing only in males.

A final point regarding the interplay of development and evolution. An age-specific approach to ontogeny is discussed in Chapter 1. In the language of this section, an age-specific approach would lead us to treat different developmental stages as involving 'complete' sets of behavioral mechanisms, rather than simply mechanisms in-the-making for adulthood. The idea of age specificity contributes to the integration of evolution with ontogeny. If each developmental stage is complete and so can stand on its own, then each constitutes a functional phenotype available to be naturally selected, should the ecological demand arise. The process through which this would occur is called *heterochrony*, i.e., acceleration or retardation of some development changes relative to others (Gould, 1977). The richness of the raw material available in ontogeny for evolutionary change becomes clearest when one recognizes the fact of age specificity of behavior.

2.5 The interplay between assessment and management in multiple time frames

The consequences of communicative behavior feed back in many time frames, including the five discussed above – immediate, tonic, developmental, short-term evolutionary, and long-term evolutionary. From a regulatory perspective, we expect such feedback to be selective in all time frames, tending to preserve those ways of behaving that yield positive consequences. Keep in mind that the definition of 'positive consequence' may vary, depending in part on the particular time frame of interest. Immediately positive effects, for example, might involve sensory gratification or reduction of aversive inputs. An individual might stop an attack on itself, for example, with an appeasement signal. Evolutionarily positive effects, on the other hand, have to do with fitness consequences. Of course, these two are linked; the extent to which an immediate effect is judged positive is influenced by its fitness consequences.

In Chapter 1, some limited evidence is cited that proximate feedback is important for management. The social consequences of singing affect the development of song structure in white-crowned sparrows and brown-headed cowbirds. And, the social consequences of antipredator calling by infant vervet monkeys may have significant immediate and developmental effects on these individuals' subsequent antipredator behavior. According to an assessment/management perspective, the proximate impact of feedback on communicative behavior should be a major emphasis of future research efforts.

The evolutionary importance of feedback through natural and sexual selection is a central idea in an evolutionary approach to communication. The evidence for the importance of such feedback processes lies not only in the remarkable heuristic value of an evolutionary approach, but also in the accumulation of data on specific communication systems indicating that selection is important (see Chapter 1). Motivation–structural rules, the relationship between vertebrate vocal structure and aggressive motivation, arose through selection on vocal structure imposed by the risk assessment rules of adversaries. And, the *chuck* portions of male Túngara frog calls were favored by sexual selection arising from the response properties of the auditory systems of female Túngara frogs.

The most important source of selective consequences of management is the assessment systems of targets. This accounts for the widespread evidence that managerial activities are designed (as in Thompson, 1986; 1997) to capitalize on the assessment systems of managerial targets. Concepts such as sensory drive (Endler, 1992), sensory exploitation (Ryan, 1994), and receiver psychology (Guilford & Dawkins, 1991) have been founded on this theme. And, it was argued early in this chapter that Batesian mimicry involves exploitation by a third party of the assessment side of a dyadic communication system (Markl, 1985).

Such exploitation of assessment by management influences assessment, both proximately and ultimately, by changing the many consequences of applying particular rules of assessment. Remember motivation–structural rules, for example. The evolutionary roots for these rules were hypothesized to lie partly in the physical link between body size and the pitch of sound produced. The evolution of motivation–structural rules reduced the relationship between sound pitch and body size, substituting a relationship between vocal pitch and the social motives of the vocalizer. Adjustments in assessment rules would need to be made in response to these modifications in the correlates of voice pitch.

Of course, assessment systems do not simply respond to changes induced by management systems. The management sides of dyadic communicative systems are also potentially exploitable resources. When third parties use assessment to exploit the signals of a dyadic system of communication, the process is called eavesdropping. The fringe-lipped bats of the Prologue caught frogs by eavesdropping on the *chuck* notes of the males' vocalizations to female frogs. Eavesdropping is not just an interspecies phenomenon; nontarget conspecifics also make use of vocalizations directed at others (see Section 3.1). So, assessment is evolutionarily designed to exploit management by extracting cues from management activities, which includes adjusting for the distortions in assessment rules sometimes imposed by management.

Proximate adjustments of assessment systems are made at many levels and time frames. The learning ability called Pavlovian conditioning is widespread in vertebrates, and reflects a mechanism that mediates the rapid acquisition of new assessment cues. For example, many vertebrate species develop coherent antipredator systems as part of their species-typical repertoire. The behavioral portions of these systems are fairly limited in their plasticity; animals deploy a relatively small set of behavioral responses to predatory situations, and do not readily learn different responses (Coss & Owings, 1985; Fanselow, 1989). But, they do learn a variety of cues of impending danger through Pavlovian conditioning. Previously neutral stimuli can become cues to danger (conditioned stimuli) if paired with already painful or frightening stimuli (unconditioned stimuli). Sometimes the unconditioned stimuli are the managerial outputs of other individuals, such as the antipredator vocalizations of conspecifics or heterospecifics. The studies described in Section 2.3.2, of rhesus monkeys learning to fear snakes and European blackbirds' acquisition of fear of novel birds, provide examples of this sort. In both cases, individuals learned to treat initially neutral animals as sources of danger when they observed a conspecific directing vocal and other antipredator activities toward the neutral animal. In other cases, detection of predators is the unconditioned stimulus, and the vocalizations of heterospecifics are the conditioned stimulus. For example, golden-mantled ground squirrels respond as strongly to yellow-bellied marmot antipredator calls as they do to their own (Shriner, 1995). A field experiment that paired simulated hawk flyovers with the playback of novel sounds indicated that these squirrels may come to treat marmot calls as danger cues through a natural Pavlovian conditioning process.

2.5.1 Signal form and its link to signaling situations

In Section 2.1 we discussed the rattling sound produced by rattlesnakes that burrowing owls mimic with their hissing vocalization was discussed It was noted that rattling is an aposematic signal, that is, a sound used to ward off potential sources of injury to rattlesnakes by warning that the snake is venomous. This aposematic function is probably the source of selection that has generated the evolution of rattling by rattlesnakes

(Klauber, 1940; Greene, 1988). Nevertheless, active assessment processes by California ground squirrels have put that sound to more subtle uses (Rowe & Owings, 1978; Swaisgood, 1994; Rowe & Owings, 1996). These squirrels use the sound of rattling not only to judge that they are dealing with rattlesnakes, but also to assess the snake's body size and body temperature, two major determinants of the danger that these snakes pose to squirrels (Rowe & Owings, 1990; Swaisgood, 1994). The structure and patterning of the rattling sound have proven to be a rich source of assessment cues. These ectothermic predators are more dangerous when they are warmer, delivering strikes with less hesitance, higher velocity and greater accuracy. The rattling sound affords cues about this source of danger: warmer snakes rattle with faster click rates, higher amplitudes, and shorter latencies. Similarly, larger rattlesnakes are also more dangerous, delivering more venom per strike (Kardong, 1986), and striking with higher velocity and over longer distances. Again, this dimension of danger can be inferred from the higher amplitude and lower dominant frequencies of the rattling sounds of larger snakes. California ground squirrels use these cues (Fig. 2.11), responding with greater caution to playbacks of rattling sounds recorded from larger and warmer snakes (Swaisgood, 1994).

How did variation in the structure and patterning of rattling come to be so informative? Not through selection for more effective signal function. Variation in rattling is no richer a source of cues about differences in snake dangerousness than the nonsignaling act of striking is (Rowe & Owings, 1996; Fig. 2.12). Rattling and striking are similar in having become potential sources of assessment cues through the physical and physiological factors that constrain and account for variation in their form (Hennessy *et al.*, 1981; Owings & Hennessy, 1984). The body size and body temperature cues available in rattling have their origins in relatively straightforward physical and physiological constraints. For example, larger snakes can strike with higher velocity, and the dominant frequency of their rattling sounds is lower because their rattles are larger and therefore have lower resonant frequencies. Similarly, colder snakes strike more slowly and rattle with lower click rates because muscle-contraction speed declines as muscle temperature drops.

One may wonder why rattlesnakes should be rattling at their ground squirrel prey. The answer is that they do not. Only California ground squirrel pups are rattlesnake prey; it is typically adult California ground squirrels that confront rattlesnakes. Females with young are especially likely to spend time deflecting hunting snakes from their pups, and press-

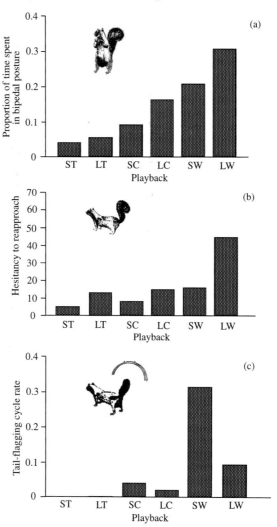

Fig. 2.11. California ground squirrels can distinguish the sizes and body tempera-tures of northern Pacific rattlesnakes on the basis of the structure of their rattling sounds. Results of a field study in which six sounds were played back: ST = soft tone (control sound); LT = loud tone (control sound); SC = rattling sound of small, cold snake; LC = rattling sound of large, cold snake; SW = rattling sound of small, warm snake; LW = rattling sound of large, warm snake. Significant comparisons for the two minutes following playback: (a) bipedal posturing: warm versus cold; (b) hesitancy: warm versus cold, and large versus small; (c) tail flagging: warm versus cold, and large versus small. The reversal of the difference between LW and SW for tail flagging may reflect the need to reduce tail flagging when being very vigilant (while hesitating and bipedal). (Graphs courtesy of Ron Swaisgood; squirrel drawings courtesy of Dick Coss.)

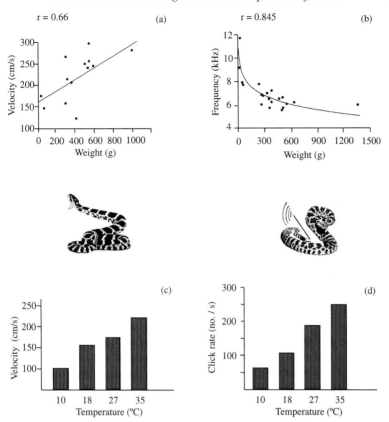

Fig. 2.12. Examples of how danger from northern Pacific rattlesnakes increases with their body size and body temperature, and of some of the cues extractable from the rattling sound of these snakes about these changes in danger. (a) Strike velocity increases with snake size. (b) Dominant spectral frequency of the rattling sound declines with snake size. (c) Strike velocity increases with body temperature. (d) Click rate of rattling increases with body temperature. (Graphs courtesy of Matt Rowe; left snake drawing courtesy of Dick Coss.).

ing the snakes enough to evoke defensive behavior, including rattling (Swaisgood, 1994). Confronting rattlesnakes might seem foolhardy, but these squirrels have evolved physiological defenses that allow them to neutralize rattlesnake venom where rattlesnake predation has been a long-time source of selection (Poran & Coss, 1990; Coss *et al.*, 1993). This adaptation protects adults from lethal rattlesnake bites, but not from injury. Pups, on the other hand, do not have enough of the venom-neutralizing substances to survive the average rattlesnake bite.

The rattling example illustrates the general answer that an A/M approach provides to questions about where the informative cues in signals come from. Structural features of signals are not arbitrarily 'chosen' to convey particular messages. Informative cues arise from the variety of constraints operative in the behaving situation. This is true even where, in contrast to rattling, variation in signal form has been shaped by its communicative function. The idea of situational constraints helps us to understand why non-signaling activities like striking by rattlesnakes are often as rich a source of assessment cues as signaling activities like rattling are. In Chapter 3, the extent to which the assessment systems of others are major sources of such constraints, where signals are involved, is discovered. The idea of assessment-imposed constraints provides a rich springboard for exploring the communicative significance of variation in the structure and patterning of signaling activities.

3

Form and function in vocal communication

3.1 Consequences, society, fighting, and the origin of vocal communication

This chapter describes how the consequences of vocalizations have affected their form, i.e., their physical structures. Social consequences of vocalizing result from assessment, but there are other consequences too. If a bird is killed by a predator attracted to its song, the predator's attack is a side-effect, not a social consequence, of singing. Therefore, conspicuous singing, and other communicative behavior, is selected for if, *on balance*, given side-effects, it results in fitness benefits. Another class of side-effects might accrue during eavesdropping by nontarget individuals (Myrberg, 1981; McGregor & Dabelsteen, 1996). For example, female lions roar to stay in contact with their pridemates and defend their territories against other prides. In doing so, they risk attracting the attention of potentially infanticidal males. Females counteract this eavesdropping by roaring in choruses, which intimidates these males (Grinnell & McComb, 1996). So, both social consequences and side-effects are sources of selection on communication.

Social consequences arise chiefly from assessment. The assessor perceives a signal. It might do nothing, it might run away, it might share, it might copulate, etc. By assessing and acting in its own interest, the assessor provides feedback to the manager to determine what has been accomplished. This process drives the evolution of communication. But what if the assessor does not act in the manager's interest at all? The only road left for the manager is to change the assessor's behavior physically by fighting and this is common enough. And here we have a hint about the possible evolutionary origin of communication – communication substitutes for fighting over resources. Resources refer to commodities outside an individual organism's body (e.g., food, mate, shelter, or simply staying

Table 3.2. *Mammalian sounds used in hostile or 'friendly,' appeasing contexts.*

Species (family)	Hostile	Friendly or appeasing	Source
Virginia opossum, *Didelphis marsupialis* (Didelphidae)	Growl	Screech	Eisenberg *et al.* (1975)
Tasmanian devil, *Sarcophilus harrisii* (Dasyuridae)	Growl	Whine	Eisenberg *et al.* (1975)
Wombat, *Vombatus lasiorhinus* (Phascolomidae)	Deep growl	*quer-quer-quer*	Eisenberg *et al.* (1975)
Guinea-pig, *Cavia porcellus* (Caviidae)	Grunt, snort	Squeak, *wheet*	Eisenberg (1974)
Mara, *Dolichotis patagonum* (Caviidae)	Low grunts	Inflected *wheet*	Eisenberg (1974)
Curo curo, *Spalacopus cyanus* (Octodontidae)	Growl	Short squeaks	Eisenberg (1974)
Degu, *Octodon degus* (Octodontidae)	Growl	Inflected squeak	Eisenberg (1974)
Spiny rat, *Proechimys semispinosus* (Echimyidae)	Growl	Twitter, whimper	Eisenberg (1974)
Agouti, *Dasyprocta punctata* (Dasyproctidae)	Growl, grunt	Squeak, *creak-squeak*	Smythe (1970)
Pocket mouse, *Heteromys* (2 species) (Heteromyidae)	Low, scratchy growl	Whining squeal	Eisenberg (1963)
Pocket mouse, *Liomys pictus* (Heteromyidae)	Low, scratchy growl	Whining squeal	Eisenberg (1963)
Desert pocket mouse, *Perognathus* (4 species) (Heteromyidae)	Low, scratchy growl	Whining squeal	Eisenberg (1963)
Kangaroo rat, *Microdipodops pallidus* (Heteromyidae)	Low, scratchy growl	Whining squeal	Eisenberg (1963)
Kangaroo rat, *Dipodomys* (6 species) (Heteromyidae)	Low, scratchy growl	Whining squeal	Eisenberg (1963)
Lemming, *Dicrostonyx groenlandicus* (Cricetidae)	Snarl, grind	Whine, peeps, squeals	Brooks & Banks (1973)
Uinta ground squirrel, *Citellus armatus* (Sciuridae)	Growl	Squeal	Balph & Balph (1966)
Maned wolf, *Chrysocyon brachyurus* (Canidae)	Growl	Whine	Kleiman (1972)
Bush dog, *Speothos venaticus* (Canidae)	Buzzing growl	Squeal	Kleiman (1972)
Coati, *Nasua narica* (Procyonidae)	Growl	Squeal	Kaufmann (1962)
Large spotted genet, *Genetta tigrina* (Viverridae)	Growl-hiss	Whine or groan	Wemmer (1976)

Species (family)	Hostile	Friendly or appeasing	Source
African elephant, *Loxodonta africana* (Elephantidae)	Roaring, rumbling sounds	High-frequency sounds	Tembrock (1968)
Indian rhinoceros, *Rhinoceros unicornis* (Rhinocerotidae)	Roaring, rumbling	Whistling	Tembrock (1968)
Pig, *Sus scrofa* (Suidae)	Growl	Squeal	Tembrock (1968)
Llama, *Lama guanacoe* (Camelidae)	Growl	Bleat (long distance only?)	Tembrock (1968)
Muntjac, *Muntiacus muntjac* (Cervidae)	Not given	Squeak	Barrette (1975)
Squirrel monkey, *Saimiri sciureus* (Cebidae)	Shriek calls, err	Peep calls, trills	Schott (1975)
Spider monkey, *Ateles geoffroyi* (Cebidae)	Growl, roar, cough	*Tee tee*, chirps, twitter, squeak	Eisenberg & Kuehn (1966)
Rhesus monkey, *Macaca mulatta* (Cercopithecidae)	Roar, growl	Screech, clear calls, squeak, nasal grunting, whine, long growl	Rowell & Hinde (1962)

3.2 is how they function to convey an impression of the size of the vocalizer.

If vocalizing is to replace size, ESS predicts low, harsh, sounds as the incontestable signals for winning aggressive encounters because there is a direct relationship between low frequency and the size of the animal producing the sound. The larger the animal, the lower the sound frequency it *can* produce. This is simply a law of the physics of sound production and resonance (compare the sound of a snare drum with that of a bass drum). Large size and low pitch provide incontestable signals for assessment. Harshness, the entity that provides the 'growl' quality to aggressive sounds, is also based on physics: a vibrating vocal membrane will produce multiple frequencies of sound as the membrane's tension lessens (Greenewalt, 1968). Consequently, the use of low-frequency sounds by aggressive animals will be accompanied by increased harshness (or 'growliness') due to the harmonic production and other off-tonal sounds produced by a vibrating flaccid membrane. Harshness, due to its physically inevitable association with low-frequency sound production, has been ritualized as a component of aggressive communication. This means that harsh sound quality is no longer simply a consequence of low-frequency sound production but may be incorporated, independently, into higher-pitched vocalizations. In this way, managers can threaten assessors with a high degree of fine-tuning to the assessment process.

In contrast, *nonvocal* sounds may or may not reflect body size and are usually not related to ESS. Nonvocal sounds are those not produced with the vocal organs. Vocal sounds are defined as sounds originating from organs whose sole purpose is to produce sound, such as the syrinx in birds and the larynx in mammals. The origin of sounds from these specialized organs involves the central nervous system, which ties vocal sound production with motivational state. This tie makes it difficult to produce vocalizations that do not reflect motivation 'truthfully.' Because animals are poor actors, ESS is inevitably 'honest signaling' (i.e., incontestable). As such, the truthful expression of motivation via vocalizations can be selected *against* in favor of nonmotivationally based communication. For example, when the assessor is larger than, and/or a predator of, the manager, nonvocal sounds such as hisses, stridulations, and lip-smacking are used because *a vocal indication of motivation would not result in effective management.* Think, for example, of predator–prey interactions: a tiny chickadee (*Parus* sp.) would gain little by growling at a squirrel at the entrance to its nest hole, even though chickadees 'growl' at one another. But chickadees do hiss, perhaps mimicking a snake, to threaten effectively (Pickens, 1928; Gompertz, 1967; Rowe *et al.*, 1986).

Amphibians and reptiles, that grow throughout life, provide a direct test of the heuristic value of ESS, for their vocalizations should directly reflect size – larger individuals should produce lower frequencies than smaller ones. This relation has been shown repeatedly in amphibians (Davies & Halliday, 1978; Ryan, 1980; Sullivan, 1982; Arak, 1983). The correlation between 'deep croaks' and large size is well established. It is necessary to confirm that this correlation is ESS by determining whether the sound can function apart from the actual body size we are suggesting is being 'symbolized.' Davies and Halliday (1978) did this by showing that an artificial playback of a larger toad's call repels a smaller toad. The sound *alone* is sufficient to produce the effect 'desired' by the manager. We suggest that ESS is the basis for incontestable signals and evolved due to this form of A/M. The evolutionary origins of vocal communication suggest that the simple politics of toad society may still be found in human politics, as when diplomatic discussions between nations forestall war.

But more complicated use of communication arose when the direct size symbolism of amphibian vocalizations was augmented with motivation in warm-blooded vertebrates. Birds and mammals, as endotherms, have high metabolism compared to other vertebrates and it is no accident that their size at sexual maturity remains relatively stable. High metabolism increases the cost of large size. Body sizes became stable at the largest size conducive to reproductive success, usually larger for males than females in endothermic vertebrates. Endothermy led to complex social behavior and parental care which, in turn, added kin selection and selection for 'Machiavellian' intelligence to prosper in complex social groups (Humphrey, 1976; see also West-Eberhard, above). These factors racheted up the number and variety of vocal signals favored by selection. Management was no longer a simple matter of 'sounding big,' because all individuals were similar in size.

Actually, ESS probably took on more complicated social functions as long ago as the Cretaceous, 60–130 million years ago, before birds and mammals were dominant but perhaps still associated with endothermy. Lambeosaurine dinosaurs composed a subfamily of the hadrosaurs, the most abundant and diverse group of large terrestrial vertebrates of the Northern Hemisphere in the Late Cretaceous. Hopson (1975) hypothesized that the crests adorning these dinosaurs functioned as acoustic resonators. Fossil evidence suggests rather intricate social behavior, including close parent–offspring ties (Hopson, 1977; Currie & Sarjeant, 1979; Horner & Makela, 1979). Weishampel (1981) used resonance analysis and auditory anatomy to analyze these fossil crests for their potential as acoustic resonators. He suggested that lambeosaurines vocalized over a wide but predominantly low range of frequencies in adults and that sexual dimorphism in vocalizations was present in some species. Juveniles vocalized at higher frequencies than adults, and potential auditory sensitivity at high frequencies in adults suggested a high degree of parent–offspring vocal communication, similar to that still found in crocodilians.

All this suggests that ESS was probably well established in these precursors of today's birds and mammals. Oddly, vocal communication apparently was more widespread in dinosaurs than it is now in reptiles, as geckos, crocodilians, and a few anoles are the only living reptiles that vocalize, although many reptiles hiss, stridulate, or rattle (Bogert, 1960). Perhaps vocal behavior was extensive in dinosaurs because some were warm blooded (Folger, 1993) and, therefore, more energy restricted than most of their cold-blooded descendants? If so, this would favor the management of others' behavior via communication.

3.3 The motivation–structural code in birds and mammals: beyond size symbolism

The sounds used by aggressive birds and mammals are low in frequency, whereas fearful or appeasing individuals use high-frequency sounds (see Tables 3.1 and 3.2). Quickly scan the sounds listed under 'aggressive' and imagine them occurring simultaneously. Low, harsh sounds are consistently associated with aggressive motivation. A similar relationship between higher, tonal sounds and friendly or appeasing motivation is found if the sounds listed in the 'nonaggressive' column are scanned. We are also able to appreciate the convergence in underlying motivation, because humans use vocal intonations expressing aggression or appeasement in the same general way (Ohala, 1980; 1984).

Unlike communication in amphibians and reptiles, however, avian and mammalian vocalizations reflect differences in *motivation,* the same individual producing a range of vocal frequencies and even combining different signals into compound vocalizations.

In addition to listening to them, the validity of ESS can be observed. Birds and mammals often increase or decrease their apparent size by postures, elevating or depressing feathers or hair coincident with production of higher or lower vocalizations, respectively. It is no accident that a growling dog erects its fur and arches its back, while the same dog crouches and sleeks its fur when it whines. These postures were noted by Darwin (1872/1965), who then framed the 'principle of antithesis,' that opposite postures are used to depict opposite moods so as not to confuse the message. Collias (1963) applied this principle to vocalizations.

3.4 The motivation–structural code

The M–S code provides an empirically based theory with the power to be directly tested on living animals and falsified. If motivational control were not based on the M–S code, many examples not fitting its predictions for sound structures should exist. The M–S code could be ancient, as we suggest, or it could evolve *de novo* thousands of separate times when different species undergo similar selection pressures. If the code is ancient, then its genetic basis will be homologous throughout the terrestrial vertebrates. It seems highly unlikely that any widely held biological trait developed *de novo* in each species. Of course a trait can be selected against and disappear from certain species (e.g., adult New World vultures are mute). The lack of functional eyes in a wide taxonomic range of animals living in lightless caves is an example of the loss of a trait through selection pressure acting against genes for eye development. Because the close relatives of these cave animals have functional eyes, we say that eyes are lost in the cave species, not that eyes arose independently in all the noncave species.

Figure 3.1 is a diagram of the M–S code showing intermediate structures between the two motivational 'endpoints' (lowest harsh and highest tonal) as illustrated in Tables 3.1 and 3.2. Figure 3.2 shows the motivational code overlain on the structure code that reflects ESS. Each block in Figure 3.1 shows a sound with two dimensions, frequency (height above the base) and relative harshness (thick line) or tonality (thin line). These two dimensions can vary independently, as can visual displays (Wiley, 1975), to provide more management power, but lower sound frequency generally coincides with wider bandwidth, or harshness, as described in Section 3.2. A third and forth dimension to the M–S code, not depicted, are the rate of delivery and the loudness of the vocalization. Increases in rate and loudness function like exclamation marks and emphasize the urgency of the situation.

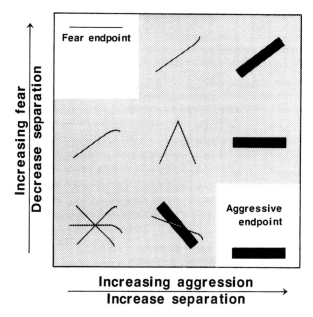

Fig. 3.1. A diagrammatic representation of vocal forms to illustrate the motivation–structural code. Each figure shows a hypothetical sound spectrogram: thin lines mean a tone or whistle-like sound, and thick lines harsh or broad-band sounds. The vocal forms grade from high pitched and tonal (fear endpoint) to low pitched and harsh (aggressive endpoint). In the lower left figure, motivation is weakly tending toward aggression if the thin line slopes downwards (its pitch drops towards the aggressive endpoint), and toward fear if the line slopes upwards. In the center left vocal form, closer to the fear endpoint, vocalizations rise variously upwards and are tonal. The three figures on the right (aggressive) side of the diagram are all broad band, but the pitch is rising in the 'distress' call (upper right corner), where fear and aggression are evoked with equal strength. The central form depicts a chevron (the 'bark') because the motivation and vocal form are squarely between the endpoints and therefore the form both rises and falls in pitch.

The *code* in the M–S rules is the decipherable grading of acoustic form from one endpoint to the other endpoint. Lower and harsher or higher and more tonal vocalizations show motivational tendencies towards opposite ESS endpoints. The vocalization within each cell (Fig. 3.1) can rise or fall in frequency range and change in quality to grade into adjacent cells. 'Fear' and appeasing states, while different motivations, are lumped together because ESS predicts that 'smallness' is important to both. Symbolizing small size works to convince the assessor to remain close, allow the manager to get closer, or reduce the likelihood that the assessor will attack.

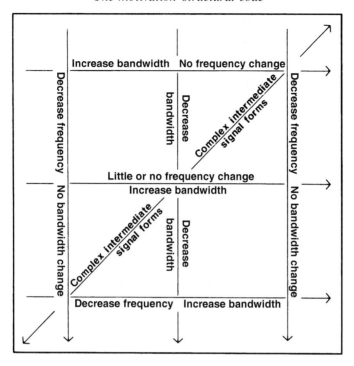

Fig. 3.2. The same diagram of motivation–structural rules but highlighting how each vocal form seen in Fig. 3.1 may vary to code changes and gradations in motivation.

Scream vocalizations, often called distress calls, are depicted in the upper right cell. When both fear and aggression are conflicting, the calls are both harsh and rising in frequency. As predicted by these component structures, the caller may attack or flee (the 'cornered rat' syndrome). This acoustic form often lacks species-specific attributes because it often occurs between species. Captured prey use distress screams as a last-ditch effort to manage the predator that holds them. Their use is adaptive because they may startle a predator, bring on mobbing, or attract a larger predator to compete for them (Driver & Humphries, 1969; Hogstedt, 1983). This last-ditch effort at management must succeed, at least rarely, because distress screams are common. The distress code type has been studied by Aubin (1987; 1991), who confirmed experimentally the interspecific nature of the distress call. Although screams often lack species distinctiveness, many other vocal forms are species specific even though the code underlying them is interspecific or general.

The code forms a 'backbone' of basic frequency and sound qualities over which a species' distinctive veneer is laid.

In the center of Figure 3.1 is a code type rendered onomatopoetically as a *bark* in descriptions of mammal vocalizations (or as a *chip* in small birds; we will use *bark* as a general term for convenience). *Barks* are characterized by a rise and fall in pitch (or fall and rise) and they are often short and abrupt. Whereas screams emanate from an animal that is simultaneously very frightened and aggressive, barks are completely intermediate to the fear and aggression endpoints. However, an individual can raise or lower the overall pitch of its *barks*, tending towards an endpoint. Their chevron form encodes this intermediate position for they do not consistently rise towards the fear endpoint nor do they lower towards aggressive. *Barks* are elicited by stimuli that do not warrant attacking or withdrawing from, but at the same time are important stimuli to manage. The generic term 'stimuli' is used here because they range from conspecific competitors for territory, group mates that might stray away from one ('contact notes'), to predators that are not dangerous once noticed. For example, a dog will bark from within its house when it hears someone at the door but then growl (bristling its fur to appear larger) at a stranger or produce high-pitched whines (and sleek its fur) if the person is familiar. Again, birds mobbing a perched owl or hawk often bark and repeatedly approach and withdraw from it, but they do not attack such potentially dangerous predators. *Barks* are often used by a prey species for predator mobbing because direct aggression by a prey species against a hawk is precluded but the stimulus is important. Vocalizing in this context has been favored by selection to make the predator move away (it can no longer surprise prey), to safeguard kin, to reduce the vocalizer's vulnerability, and other reasons. Such observations suggest the hypothesis that the *bark* by itself codes neutral or adaptively indecisive motivation. *Barks* function in a great diversity of contexts because a manager can either attract or repel assessors. To assessors, barks say 'I discovered a stimulus of interest to me and it is probably of interest to you,' or 'I am here and I am the same species as you are.' Many migratory birds defend nonbreeding territories using species-distinctive barks in place of song because barks alert both sexes to the manager's presence.

A particular acoustic structure has a particular motivational significance and function. This limits the range of situations in which that structure is useful for management. Assessment may eliminate some acoustic form categories in some species, and many categories may be missing from a species-typical vocal repertoire such as when visual communication replaces vocalizations in situations of close contact or open habitats.

The M–S code depicts a fluid and interconnected, not a stereotyped, series of acoustic forms. In addition to grading among categories (or displays), various 'compound' vocalizations consist of two or more code types. For example, the *chickadee* call of the black-capped chickadee (*Parus atricapillus*) consists of two to four chevron-shaped notes ('chicka') that decrease in frequency, followed by still lower-pitched harsh 'dee' sounds (Ficken, Ficken & Witkin, 1978; see Fig. 3.3). These components are also used separately alone, which is why the call is defined as being compound. Using the M–S code, the sound might be described in general terms as a series of barks followed by low harsh sounds. The *barks* suggest the hypothesis that the manager has perceived something of interest, with the decrease in their frequency suggesting increasing aggressiveness. The ending 'dee' notes are aggressive but not at the aggressive endpoint for this species. Even lower and harsher sounds are used in fighting. The stimulus eliciting this compound call is of interest (middle of the code chevrons) but is not causing fear or escape, as suggested by the broad-band sound at the end ('dees'). Using sound structure to predict managerial function, it may benefit the manager to attract other chickadees in the presence of stimuli that elicit a compound sound with these motivation–structural components. Fieldwork has shown conspecific attraction to be a function of the *chickadee* call (Ficken *et al.*, 1978).

Acoustic forms and their placement within the code matrix (see Fig. 3.1) allow one to use the information in a spectrogram to produce testable hypotheses about a signal's function. For example, the Carolina wren (*Thryothorus ludovicianus*) has 11 categories of close-range signals, seven grading extensively or combined to form composite signals (Table 3.3). Let us focus on one of these, the *chirt*, used for predator surveillance and intraspecific interactions (Morton & Shalter, 1977). Some natural history of this species, important to understanding signaling, is described in the Prologue. In the presence of a hawk, the caller presents *chirts* in bouts of two to five at the rate of around seven per second. The second *chirt* in a bout is higher in pitch than either the first or subsequent *chirts* in the bout. This changes abruptly if the hawk makes the slightest move: the *chirts* remain at the higher pitch of the second *chirt;* if the caller moves to escape, the *chirts* rise in pitch even more (Fig. 3.4). In intraspecific use, *chirts* are lower in pitch and harsh when the caller has chased another

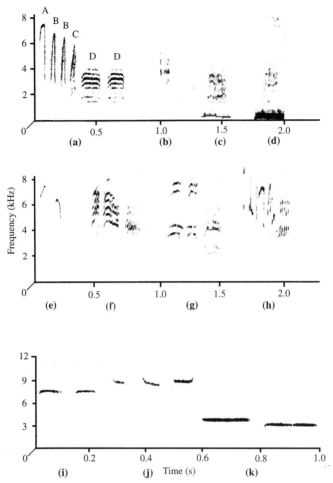

Fig. 3.3. Sound spectrograms of black-capped chickadee vocalizations: (a) chick-a-dee call; (b) twitter; (c) hiss; (d) snarl; (e) tseets; (f) begging dee; (g) broken dee; (h) gargle; (i) high zees; (j) variable sees; (k) fee-bee (song). Part (a) illustrates a compound vocal structure consisting of four types of notes (A, B, C, D). Except for the song (k) used in long-distance communication, the vocal forms fit the predictions of motivation and function predicted by M–S rules. (After Ficken *et al.*, 1978.)

wren from its territory, or higher and more tonal when it is contacting and attracting its mate (Fig. 3.5). When the mate approaches, the *chirt* combines with *pizeet,* a signal used in close mate contact and appeasement, to form a compound signal.

The code operates in the *chirt* system in the following way. First, *chirts* are a form of *bark,* fitting (in Fig. 3.1) between the *bark* and the upward

Table 3.3. *Vocalizations of the Carolina wren (*Thryothorus ludovicianus*) arranged in order of decreasing distance between communicants.*

Vocalization	Distance between communicants	Acoustic structure/ variation
Male song	Long	Tonal/none
Male *cheer*	Long to medium	Tonal/none
Female *chatter*	Medium to short	Harsh/little
Female *dit*	Medium to short	Tonal/little
Rasp	Short	Harsh/great
Chirt	Short to medium	Tonal to harsh/great
Pi-zeet	Short	Tonal to harsh/great
Male *tsuck*	Short	Tonal/little
Female *nyerk*	Short	Harsh/moderate
Scee	Short	Tonal to harsh/great
Pee	Short	Tonal to harsh/great
Growl	Short	Harsh/little

tonal or level harsh cells, depending on context, as described in the last paragraph. The Carolina wren's *chirt* is not at all like mobbing and the birds do not approach the hawk. *Chirts* are a unique form of hawk surveillance, permitting wrens to forage in the presence of a perched hawk. The rising and sustained higher pitch evoked by a moving hawk is correlated with a higher likelihood of the hawk attacking. Fear increases, as does the pitch of the *chirts*. Generally, *barks* are favored in mobbing and in the *chirt* surveillance system of the Carolina wren, but small birds use high-pitched and tonal vocalizations when a hawk surprises them – the fear endpoint (Box 3.1).

Female Carolina wrens cannot defend a territory alone; their survival during the winter depends on long-distance signaling (song, *cheers*) by the male. Therefore, it is no surprise that females do most of the hawk surveillance through *chirts*. By chirting to keep the male alerted, and by providing the male with time to forage in the presence of an otherwise dangerous hawk, he is more likely to survive deep snow so as to protect the female's territory (Morton & Shalter, 1977). The evolution of this vocal predator surveillance system is entirely based on the self-interest of the birds. In this case, the self-interest of both pair members coincides, resulting in a stable communication system without conflict between manager and assessor. A mammalian example of predator vigilance is found in the dwarf mongoose (*Helogale undulata*). Rasa (1986) describes

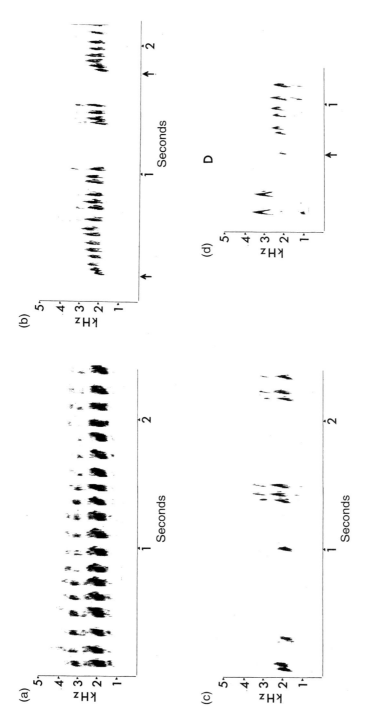

Fig. 3.4. Sound spectrograms of the *chirt* system of the Carolina wren. (a) Evenly spaced, rapid *chirts* typical of those uttered during reaction to a dangerous, moving hawk. (b) *Chirts* following two hawk head movements (arrows). (c) *Chirts* produced in variable groups, the last note drops back to the pitch of first note during periods when the hawk is still. (d) Two *barks* followed by *chirts* elicited by a hawk head movement (at the arrow). (After Morton & Shalter, 1977.)

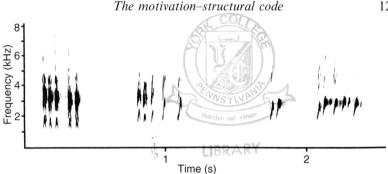

Fig. 3.5. Harsh Carolina wren *chirts* grading into tonal *chirts* produced by a wren that had just chased an intruder and then joined its mate, illustrating the overlay of motivation on vocal form within a single call type or display.

Box 3.1. *Adaptation of signals: constraints and signal structure.*

If the structure of signals encodes the constraints of their typical situations, then unrelated species experiencing similar constraints might be expected to converge in signal structure, and therefore in signal meaning to assessors. An example is the structural differentiation of *seeet* and mobbing calls shared by central European passerines of several families (Marler, 1955; Marler, 1959). The constraints of predatory situations apparently are consistent enough that even some mammalian species have converged upon similar patterns of call differentiation, e.g., as described in squirrels (Melchior, 1971; see also Vencl, 1977; Owings & Virginia, 1978; Macedonia & Evans, 1993; Tamura, 1995).

Marler's original explanation was that auditory specializations for sound localization by potential assessors in these predatory situations have shaped the structure of these calls. *Seeets* are ventriloquial sounds used when callers should minimize their locatability, whereas mobbing calls are structured for maximum localizability, e.g., to facilitate attraction of 'help' in mobbing. Although some experimental results have been consistent with this hypothesis (Brown, 1982), others have not (Shalter & Schleidt, 1977; Shalter, 1978). A study of a specific predator–prey system also indicated that *seeets* minimize the caller's conspicuousness to the predator, but do so via a different route (Klump, Kretzschmar & Curio, 1986). The frequency of great tit *seeet* calls is high enough to be much more audible to conspecifics than to hawks, whose audiograms rolled off at lower frequencies than that of the great tit. A third nonexclusive alternative is that this structural difference between *seeet* and mobbing calls reflects the constraints summarized by the motivation/structural rules hypothesis. Signals used in mobbing are *barks* whose rise then fall in frequency reflects indecision about attack or retreat. This signal is often motivated by a stimulus of great salience to the signaler and is salient to others and, therefore, is attractive to them. In contrast, the high-tonal sounds associated with evasion of dangerous aerial predators reflect the signaler's fear endpoint, which causes listening birds to become immobile.

coordinated vigilance by guards, usually subadult males, whose barks permit the rest of the group to forage intently for insects in grass tussocks. If a predator is sighted, the guards warn the group using a variety of vocalizations depending on the predator sighted (Meier, Rasa & Scheich, 1983).

In Carolina wrens, females bark with single isolated notes, or doublets and triplets when agitated, whereas males use a call that sounds like 'cheeer,' a rapid series of descending chevrons adapted for long-distance signaling (Fig. 3.6e). A Carolina wren barks at other wrens when they encroach upon its territory and when a predator is detected. Like other species, a barking wren does not attack or flee but shows interest and arousal. Males sometimes bark at other males that sing a song not in their repertoire. Wrens growl in two contexts, a stylized *growl* used only when attacking other wrens, and a harsh repeated rasping when mobbing intruding wrens or some predators such as snakes and mammals. Sometimes small mammals and snakes are attacked by rasping wrens but not by barking ones. A pattern emerges in usage: the same vocal forms used to manage predators (*chirts, rasps, barks*) are used to manage competitors when territorial integrity is challenged. The same vocal form/ motivation functions in both contexts, when ousting a predator *or* competitor is the desired management objective, because the two contexts are similar in terms of their effects upon the manager's reproductive success.

The Carolina wren is typical of many species where the M–S code is followed both within and between vocalization classes. These classes (see Table 3.3) show relatively discrete sounds that would be termed displays by earlier ethologists. The *growl* or *rasp, pee,* and *dit,* for example, are sound classes whose structures fit the aggressive endpoint, fear endpoint, and chevron-shaped *bark,* respectively, as modeled in Figure 3.1, and as illustrated in Figure 3.6e. Within-class variation is illustrated in Figure 3.7. Here, *scees* (see Table 3.3) showing acoustic structural variations are being uttered by a male Carolina wren losing in a fight with another male,

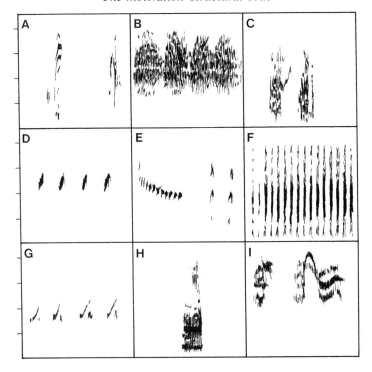

Fig. 3.6. Vocalizations of the Carolina wren inserted into the motivation–structure code matrix (Fig. 3.1) to show the close fit. Note the tonal qualities of squares A–G and the increasingly lower pitch and harshness as you move right and up. The endpoints are the growl (C) and *pees* (G), although those depicted are not as high pitched as this vocalization can become. The *pi-zeet* (A) to *rasp* (B) to growl (C) function to keep close together with mate or fledged brood mates, mob ground predators, and attack conspecifics, respectively. The barks (E) show the male *cheer*, a series of chevrons and the female *dit*, both given whenever anything of interest is perceived but when neither attack nor fleeing is likely. Females only produce *nyerk* (H) and chatter (F). Both sexes produce *scees* (I) when loosing a fight (see also Fig. 3.7). *Chirts* are illustrated in D, and were discussed earlier. See Table 3.3 for further illustrations of this species vocal repertoire.

who remained silent (Fig. 3.7). During the first 12 seconds of the fight the losing male tried to escape but was held by the winning male. The sounds during this time have a rising frequency and harsh quality, as in Figure 3.1 (upper right block), modeling a high level of both fear and aggressive motivation (i.e., distress). At 20 seconds, the sounds do not consistently rise in frequency, and at 22.5 seconds the losing male pecks back. At precisely that time, the sound it uttered became greater in bandwidth

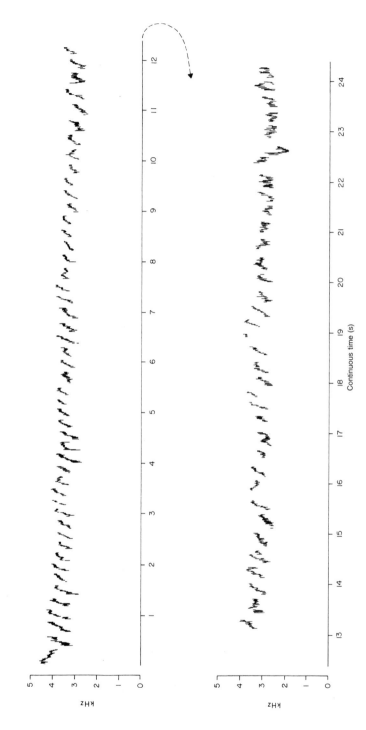

Fig. 3.7. A graded sequence of *scees* produced by a Carolina wren being defeated by another in a fight. The sharp drop and then rise in frequency near 22.5 s occurred when the bird pecked at his assailant.

and lower in frequency. These immediate changes in sound structure are correlated with behaviors consistent with aggressive or fearful motivation, a test of predictions from the M–S code. Species differ in the number of major types or classes of vocalizations and amount of gradation among and between them. Identifying selection pressures favoring the presence or absence of categories of sound structures, from the possibilities predicted in the M–S code, is an untouched area of research.

3.5 Ontogeny, asymmetries, and the motivation–structural code

The M–S code is represented in the development, use, and loss of particular vocalizations, or even whole categories, within the lifespan of individual animals. Vocalizations of smaller and younger animals are usually higher in frequency than adult vocalizations. Sibling competition may select for variable sounds, including aggressive ones (Fig. 3.8). High pitch is not inevitably associated with small body size. For example, adult-size young male mallards still *peep* like young ducklings, while their sisters have already developed the adult female *quack*. Young males retain juvenile calls longer than females because adult male–male aggression is high, and young males appear to escape adult belligerence by continuing to 'sound young.'

An example of an adaptive, age-specific, use of a vocal category, the *growl,* is found in the herring gull (*Larus argentatus*). During their first

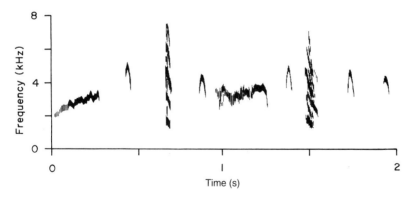

Fig. 3.8. Call notes of a three-day-old black crake chick, a precocial bird with siblings with which to compete for food, showing three structurally diverse vocalizations. The first is rising (distance decreasing) as the female approached it with food, then a bark, then a slightly aggressive downsloping harsh sound as a sibling approached.

3.7 Long-distance signals and communication

Signals broadcast to individuals that are often out of view are described as long-distance signals. It is a relative, not absolute, description since some signals function over various distances. For example, a *bark* may be used by an animal mobbing a predator, with both predator and other mobbers in close proximity, or when the same animal defends a territory against distant rivals. However, many signals (e.g., most birdsong) are almost always used for communication over long distances. Biologists study them as a class of communication apart from 'close-contact' signals because traveling long distances through the environment causes changes in the physical structure of signals (Morton, 1975). Long distances impose new constraints on A/M. For one thing, motivational state is incorporated into signal form less often than adaptations to preserve signal form during propagation. Assessors are more restricted to information extracted from signals alone, rather than from context. Cues about energetic state, RHP, mated status, and so forth, are more likely to be assessed from the signal itself and time spent signaling rather than direct assessment of the manager. Managers, in turn, must use those vocal dimensions in their signaling in order to have an effect on assessors. Common results of long-distance assessment are long bouts of singing in birds, howling in monkeys and wolves, and roaring in lions and red deer. Also, long-distance communication often involves more than two individuals at the same time. Communication among several territorial birds, involving individuals directly and indirectly with the manager, are common. Some of these individuals may be unintended targets, eavesdroppers, whose assessment activities can produce side-effects that further affect management.

For example, recall the Túngara frog in the Prologue (from Rand & Ryan, 1981). The males produce calls of varying complexity. They range from a simple single whine to more complex calls, involving whines followed by one to three chucks. Males increase vocal complexity in response to vocal competition from other males for females. Females are attracted preferentially to complex calls and to those given by larger males. One aspect of the *chuck* signal that identifies large males is its lower frequency relative to the *chuck* given by smaller males (Ryan, 1980). So, why don't all males always use the most complex call, the one designed to attract females? The answer lay in predators, especially the bat *Trachops cirrhosus,* that homes in on a frog using the complex *chuck* additions (Ryan, Tuttle & Rand, 1982; see Fig. P.4 in the

Prologue). The simple to complex call variation in this small frog, Rand and Ryan suggest, allows a male to adjust call complexity to effect a compromise between maximizing mate attraction and minimizing predation risk.

The distance an animal is able to communicate by sound depends on the source amplitude, the penetrating capability of the sound in the environment, the level of masking ambient sound or noise, and the auditory sensitivity of the individual receiving the sound. Playbacks of natural and synthetic sounds are used to study all three areas in natural habitats. One of the earlier hypotheses about signal design is that the acoustics of environments differ in ways that might affect the long-distance signals of animals that live in them (Fig. 3.9; Morton, 1970; 1975; 1986; Chappuis, 1971; Marten, Quine & Marler, 1977; Waser & Waser, 1977; Michelsen, 1978; Wiley & Richards, 1978; Bowman, 1979; Brenowitz, 1982; Cosens & Falls, 1984a; 1984b). These studies assumed that the greater the distance a vocalization was detectable, the better it was designed for its biological function(s). As will be seen later in the birdsong evolution example, detectability is not as important as signal degradation in management; signal detection, however, remains popular with biologists intrigued by engineering approaches (e.g., Klump, 1996).

Not all of the ambient sound energy (all the sound energy at a particular place, both biological and nonbiological in origin) reduces the 'conspicuousness,' or masks, a given signal. The masking effect is caused by a band of frequencies on either side of and including the signal frequency called the critical band (Fletcher & Munson, 1937). This is a frequency bandwidth such that a further widening does not increase the masking of a pure tone at its center (Bilger & Hirsh, 1956). The critical band is not an absolute entity but depends on the hearing physiology of the receiver. In general, however, the higher the signal's frequency, the wider the bandwidth of sound that masks it. Masking takes place on the basilar membrane of the inner ear when the appropriate hair cells are stimulated by the ambient frequencies to the same extent as by the signal. It is only within the critical band that ambient sound power adds to the threshold of hearing (for a particular frequency) (Fig. 3.10; Greenwood, 1961). Acoustic interference includes either partial or total masking, a physiological effect during peripheral perception of sound, and psychological effects, occurring in the central nervous system (summarized in Dooling, 1982).

Conspicuousness against noise is a complex phenomenon that depends upon the ambient noise and the hearing physiology of the listener, both

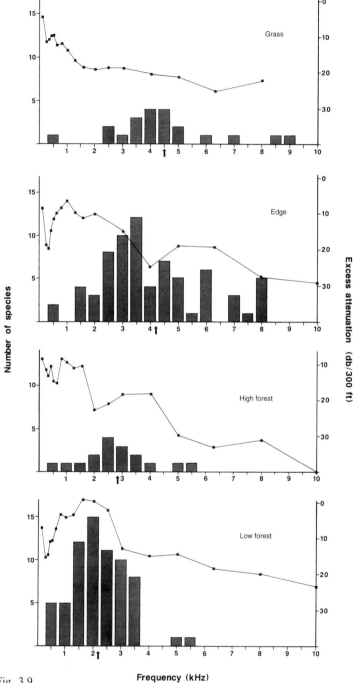

Fig. 3.9.

Lowest threshold
(db SPL re 20 μNm^{-2})

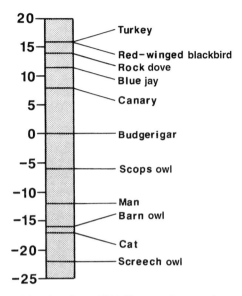

Fig. 3.10. Thresholds of hearing for a 1000-Hz tone for a variety of vertebrates, showing how variable the hearing threshold can be. Note that nocturnal animals, owls and cat, have very low thresholds, perhaps because they rely more on hearing for prey capture.

of which can affect A/M and the evolution of communication (Fig. 3.11). Species may differ in the morphology of their basilar membranes (peripheral nervous system) and psychology (central nervous system, i.e., the brain) and therefore differ in masking effects and critical bands. Frogs, for example, have ears tuned to the frequency components of conspecific calls (e.g., Capranica, Frishkopf & Nevo, 1973). The measurement of sound attenuation in natural environments involves a simple playback procedure but the results are derived from complex variables, too complex to be adequately modeled (Martin, 1981; Roberts, Hunter &

Fig. 3.9. The relation between the dominant frequencies in birdsongs and sound propagation. The number of species (black bars) using various frequencies is plotted with the excess attenuation (EA) of those frequencies in four acoustic environments: grassland, edge, high forest (canopy), and near the ground in forest (low forest). A close fit exists between EA and the frequencies forest birds use in their songs, but the fit is not as good for the open habitats of edge and grassland. Forest birds also use pure tones more than do species in the other habitats (see Table 3.4). (From Morton, 1975.)

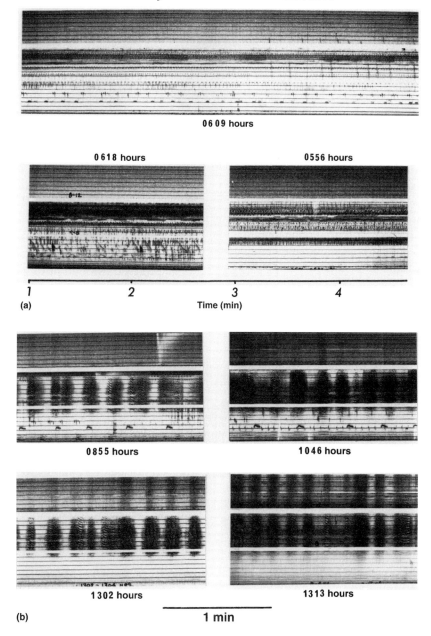

Fig. 3.11. Portions of real-time spectrograms showing the diurnal change in bio-logically produced sounds in a tropical rain forest (recorded on Barro Colorado Island, Panama). The sounds were recorded through four microphones hung under a protective shield about 9 meters above the ground. Time is indicated along the horizontal axis, frequency along the vertical axis. Sound energy is

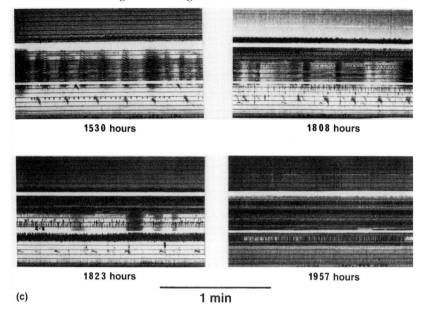

| **1530 hours** | **1808 hours** |

| **1823 hours** | **1957 hours** |

(c)   **1 min**

Fig. 3.11 (*cont.*)

indicated by darkened areas on the film. The total frequency range covered in each spectrogram is from 0 to 12 kHz. Each consists of three film strips, here shown slightly separated. The thin horizontal lines indicate 500-Hz increases in frequency going from bottom to top.

(a) 05.56 hours. Nocturnal frogs, *Eleutherodactylus diastema* (3 to 4 kHz), and insects (largely from 4 to 8 kHz) make up most of the predawn sound. A bird, slaty antshrike (*Thamnophilus punctatus*), began calling at 1.5 to 2.0 kHz toward the middle of the recording.

06.09 hours. Near daybreak, several bird species are calling: slaty antshrike, *Thamnophilus punctatus* (1.5 to 2.0 kHz), chestnut-mandibled toucan, *Ramphastos swainsonii* (centered on 2.5 kHz), and alarm notes from checker-throated antwren, *Myrmotherula fulviventris*, beginning in the last third of the spectrogram (4 to 9 kHz). Insect sounds are mainly restricted to the 4 to 8 kHz range.

06.18 hours. At least seven species of birds calling, all are below 4 kHz: slaty antshrike (*Thamnophilus punctatus*), massena trogon (*Trogon massena*), spectacled antpitta (*Hylopezus perspicillatus*), buff-throated woodcreeper (*Xiphorhynchus guttatus*), black-striped woodcreeper (*X. lachrymosus*), *Myrmotherula fulviventris*, and rufous motmot (*Barypthengus ruficapillus*). Insects are much in evidence at 4 to 8 kHz – the frogs have stopped calling.

(b) 08.55 hours. Cicadas begin calling, 3.5 to 8 kHz +, and most bird species have stopped calling except for *Hylopezus* and *Rhamphastos swainsonii*, 1.5 to 2.0 kHz and 2.5 kHz respectively.

10.46–13.13 hours. Cicadas calling, few other species are vocal. 10.46 hours has *Hylopezus* and *Thamnophilus* at 1.5 to 2.0 kHz and dot-winged antwren (*Microrhopias quixensis*) call notes at 3–4 kHz.

Kacelnik, 1981). If nothing is in the way, sound spreads spherically, its energy declining as it is spread out over the enlarging sphere formed by the advancing sound front at a rate of 6 dB per doubling of distance. (The sound will decrease at the rate of 6 dB per doubling of distance from the source to point of measurement since the density of energy at a distance r from the source decreases with the surface area of a sphere with radius r, intensity varies inversely with r^2, therefore, $10 \log (2r/r)^2 = 6$ dB). This provides a reference level for evaluating other effects (bringing about so-called excess attenuation, EA) that occur in a natural environment (Fig. 3.12). Excess attenuation is thus a comparison of the attenuation observed in a natural habitat minus that due to spreading effects. Box 3.2 provides a short course in sound physics.

Excess attenuation is usually estimated by measuring the sound level from a speaker at a close distance and with a direct line-of-sight. Measurements taken farther away are compared to these. Consequently, if the speaker's sound level varies from test to test, or if some frequencies are produced with more amplitude than others, this is taken into account in the data reduction because we are dealing with the subtraction of energy ratios (decibels) (Morton, 1975; Marten & Marler, 1977). For instance, if you measured 56 dB for a reference level at 5 m from the speaker and it measured 35 dB at 20 m, EA would be 9 dB because 12 dB would be lost due to spreading (there are two doublings of distance) and 9 dB due to the natural world.

The importance of air homogeneity to sound propagation is, in part, due to the speed of sound relative to air temperature. The velocity of sound in air is approximately $(331.4 + 0.607C)$ m/s, where C is the temperature in degrees Celsius. Sound velocity also increases a little with increasing humidity: at 20°C the sound velocity at 100 percent relative

Fig. 3.11 (*cont.*)

(c) 15.30 hours. Cicada chorus is beginning to drop in intensity, some birds are calling: *Xiphorhynchus guttatus*, 1.5 to 2.5 kHz, and dusky-capped flycatcher (*Myiarchus tuberculifer*), 2.4 kHz.

1808 hours: The transition to nocturnal insect calling begins while a few cicadas are still calling (4–9 kHz). A single frog, *Eleutherodactyla*, begins calling (3–4 kHz) and more bird species are calling.

18.23 hours. Dusk. There is a strong frog chorus at 3–4 kHz (*Eleutherodactyla*, mainly) and nocturnal insects are more prevalent.

19.57 hours. Dark. Nocturnal insects and frogs predominate. Note that the patterns of insect noise in the 4–8-kHz range are not randomly distributed but reflect competition to be heard based upon sound-frequency separation. Those insect species with continuous calls do not overlap in frequency and appear as parallel strips on the spectrograms.

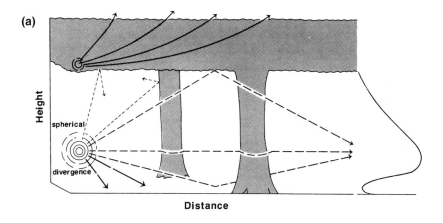

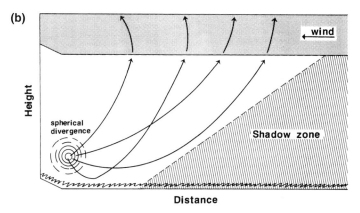

Fig. 3.12. Diagrams of how environments cause excess attenuation. (a) In the forest, near the ground, tree trunks, the ground itself, and uniform air temperature affect sound propagation. High frequencies reflect back from objects larger than their wave lengths while low frequencies are absorbed by the ground, resulting in an advantage for intermediate wave lengths called a sound window. Intermediate frequencies (e.g., 1500–2500 Hz in Panamanian rain forests) propagate better, as depicted by the convex curve on the right. Higher up in the canopy the warm air above the canopy tends to cause sound fronts to curve upwards because sound travels faster in higher air temperatures. (b) In nonforested habitats, the sun and ground plus vegetation produce distinct gradients in temperature and wind speed, respectively. Sound waves tend to bend upwards against the wind, but especially due to high-temperature gradients, producing a shadow zone near the ground where the vocalization is absent. Many grassland birds sing from high perches or in flight to overcome sound shadows.

Box 3.2. *A short review of the physics of sound propagation.*

Sound is a pressure wave which, in air, travels about 340 meters per second. The variation in pressure has two components, rate and extent. A 1000-Hz tone is perceived when the pressure changes through a complete cycle 1000 times per second, measured at a stationary point. The extent of sound pressure variation is measured in microbars; one microbar is approximately one-millionth of the normal atmospheric pressure. It may also be measured in terms of newtons per m^2 (1 newton equals 10 microbars) or in dynes (a force such that under its influence a body whose mass is 1 gram would experience an acceleration of 1 centimeter per second per second).

In sound measurements, results are stated in decibels (dB). The decibel is a common logarithm of a ratio of two values of power (e.g., watts) or sound-pressure levels (SPL). A decibel measurement may refer to a comparison of two pressure or two power levels (relative dB) or between a measured entity and a standard level (absolute dB). The standard reference levels (or '0' dB) for power is 10^{-12} watt and for SPL it is often 20 micronewtons per m^2 or 0.0002 dynes/cm^2. These measures correspond to the threshold for hearing a 1000-Hz tone for a human with average hearing ability. If one sound intensity is twice another, the number of decibels is 3 ($10 \log_{10} 2 = 3$); if one sound power is twice another, the number of decibels is 6 ($20 \log_{10} 2 = 6$). It should be remembered that sound intensity or amplitude is a physically measured entity, whereas loudness is a perceptual term. Most biologists use SPL.

humidity is about 0.3 percent greater than at 30 percent relative humidity (Michelsen, 1978). Therefore, a sound wavefront conducting through a temperature gradient will be refracted from the warmer air toward cooler air. This refraction causes a regular but fluctuating feature of open habitats called a shadow zone. The wavefront advancing parallel with the earth is defracted upward due to air temperature and wind gradients (in the upwind direction). This leaves a soundless area under the wave front, the reason one finds it so difficult to hear voices across a heated field. By contrast, if the surface cools the air immediately above it, such as over a cold lake, voices can be heard at astonishing distances. The best way for an animal in an open habitat to avoid sound shadows is to call from a perch or fly above the ground. Since the benefits of increased broadcast area increase exponentially with height, as little as 3 to 6 meters above the ground will erase most of the shadow-zone effects (Pridmore-Brown & Ingard, 1955).

Forests, especially tropical ones, have relatively homogeneous air below the canopy. Thus they differ from open habitats in lacking shadow zones for animals calling from ground level until one reaches the tree crowns. Above the tree crowns, however, the temperature gradients may again become steep. If the air within the canopy is cooler than the air above it, the canopy and ground may act as a wave guide on sounds of some frequencies emanating within the forest (Waser & Waser, 1977). These sounds will not diminish in amplitude at 6 dB per doubling of distance. As a consequence, they will experience negative EA by not spreading geometrically.

The presence and amount of reverberation constitute another major difference between open habitats, such as grasslands and forests (Morton, 1975; Richards & Wiley, 1980). Tree trunks and the forest canopy may reflect and scatter some frequencies such that the sound reaching the receiver arrives at different times. This scattering from multiple surfaces causes a decay of sharp signal onsets and offsets. The standard measure of reverberation time, RT60, is the amount of time it would take a signal to decay 60 dB after its termination. Waser and Brown (1986) found essentially no reverberation in African savanna whereas RT60s in rain and riverine forest were much longer (about 0.44 seconds) and increased rapidly and linearly with distance. Birds singing in forests have responded to reverberation and sound windows so prevalent there by narrowing the frequency ranges in their songs (Table 3.4).

Conspecifics, whose signals are likely to be of the same frequency, would be most apt to cause ambient sound that masks each other's signals. Intraspecific and interspecific communication interference is best studied in frogs, which often call in close proximity, and many species may be involved. In a series of experiments on several species of neotropical *Hyla* species, Schwartz and Wells (1983a; 1983b; 1984a; 1984b; 1985) documented both the occurrence and perception of intraspecific and interspecific behavioral responses in individual frogs. Their study of *H. ebraccata* and *H. microcephala* serves as an example. Both species call from the same microhabitats and their sounds overlap in frequency spectrum. They measured a sound pressure level (SPL) of 107 dB at 50 cm from *H. microcephala,* about 6 dB greater than the other species' call. Background noise generated by *H. microcephala* choruses causes a shift in the timing and type of calls given by nearby *H. ebraccata*. The latter reduce their calling rates and the proportion of multinote calls and aggressive calls when *H. microcephala* are calling at a high level (Schwartz & Wells, 1983a). Using playbacks of artificial sound, Schwartz and Wells determined that only a

Table 3.4. *Studies of intraspecific song variation showing narrower frequency ranges in songs from forest habitats.*

	Frequency range (Hz)		
Species	'Open' habitat	Forest	Authority[a]
Summer tanager	1362	1202	Shy (1983)
Great tit	3130	2250	Hunter & Krebs (1979)
Cardinal	3048	2408	Anderson & Conner (1985)
White-throated sparrow	830	270	Wasserman (1979)
Clay-colored sparrow	4200	2800	Knapton (1982)
Song sparrow	6274	6118	Shy & Morton (1986)[b]
Darwin's finch[c]	4500	1500	Bowman (1979)

[a]For details of references, see Morton (1977).
[b]Compared propagation over river versus field, indicating a difference in atmospheric homogeneity rather than habitat *per se*.
[c]Refers to *Camarhynchus parvulus* as an example.

3-kHz sound, the frequency that masks that of the frog's call, was effective in changing a male *H. ebraccata*'s calling. They suggest that the proximate mechanism causing the change is simple: the male *H. ebraccata* cannot hear other conspecific males because of the masking effect, and calls as though it were alone. There is a good reason for *H. ebraccata* males to avoid overlap with the louder species: female *H. ebraccata* discriminate against conspecific advertizement calls when they overlap with a chorus of *H. microcephala* (Schwartz & Wells, 1983b). In contrast to *H. ebraccata,* a third sympatric species, *H. phlebodes*, increases its calling rate and adds click notes to its advertizing calls in response to playbacks of conspecific and heterospecific calls (Schwartz & Wells, 1984a). Clearly, these small frogs make changes in their calls as adaptations to regularly occurring acoustic interference they encounter in their complex tropical communities. Note, however, that selection underlying call adaptations is generated intraspecifically through the agent of female mate choice.

Studies of anuran communities suggest strong competition for what might be called acoustic space. Hodl (1977) studied 15 species of synchronously breeding frogs that are restricted to floating islands of vegetation in the Amazon River. Most of the species (11) had mating calls differing in dominant frequency; those sharing emphasized frequency ranges within identical calling sites differed greatly in temporal features of their calls. Duellman and Pyles (1983) found patterns of total acoustic variation alike in three frog communities, but calls of closely related allopatric species were more similar than those of closely related sympatric species. Closely related sympatric species with similar calls either breed at different times or do not coexist in the same microhabitats. The adaptations of individual frogs mentioned above provide us with an evolutionary explanation of how these patterns arose.

The Puerto Rican coqui (*Eleutherodactylus coqui*) is an arboreal frog that provides an example of another dimension to the question of acoustic interference. The species is named for the males' characteristic 'Co-Qui' call. The first note, 'Co,' is a nonmodulated tone (its frequency does not change) about 100 ms long which functions in male–male territorial interactions (Narins & Capranica, 1976). The second note, 'Qui,' sweeps upward in frequency and functions to attract females. The frequency of the 'Co' call varies inversely with the size of the male. If a 'Co' call is played back to a male, he will respond by dropping the 'Qui' portion if he perceives the signal. In turn, the frog can be allowed to 'tell' one if it perceives a 'Co' played to it through the presence or absence of the 'drop the Qui' response. This is how Narins and Capranica (1976) determined that the intensity of a 'Co' playback at a calling male's own 'Co' frequency is about 30 dB lower than the level needed when he is tested with calls 200 Hz lower or higher than his own call. Neigbouring males usually do have 'Cos' of the same frequency. Since neighbors are spaced about 2 m apart, they experience their neighbors' 'Co' calls at about 83 dB SPL, roughly 30 dB above the threshold for the 'Co' note response. It follows that an individual male will not hear a neighbor whose 'Co' note differs in frequency by about 200 Hz from its own (Narins & Smith, 1986).

Mutual interference was studied in two sympatric North American bird species, the least flycatcher (*Empidonax minimus*) and the red-eyed vireo (*Vireo olivaceus*) (Ficken, Ficken & Hailman, 1974). The songs are quite different in timing but widely overlap in sound frequency. The vireo sings for long periods with about 0.2–0.4 seconds between song phrases, while the flycatcher song is short, about 0.11 seconds, and delivered infrequently. Extensive recordings of singing in five vireo–flycatcher neighbors were analyzed for overlap in songs to determine if either species was influencing the timing of song by the other. Statistical methods are provided in Box 3.3. The results showed that the flycatchers avoid beginning a song while a vireo is singing, but vireos begin songs regardless of flycatcher singing. Perhaps the flycatcher avoided the vireo because its shorter song could suffer total interference by the long vireo song but not vice versa. In another study of interspecific effects, ovenbirds

Box 3.3. *Statistical methods to study acoustic interference in birds.*

Ficken *et al.* (1974) used the following statistical method to compare the singing overlap in two species, the red-eyed vireo and the least flycatcher.

v is the total amount of time a vireo is producing sound and x is the vireo's total quiet time in the recorded sample. Then the probability of a vireo's singing is:

$$p(v) = v/(v + x)$$

and its probability of being silent is:

$$p(x) = x/(v + x)$$

If during the total time interval $(v + x)$ the flycatcher sings f total songs, begun regardless of whether the vireo is singing or not, then the predicted number of flycatcher songs (F_r) begun during vireo song should be:

$$F_r = p(v) \times f$$

and the predicted number of flycatcher songs begun during vireo silences should be:

$$F_x = p(x) \times f$$

The predicted values can be calculated and compared with the actual values found (f_v and f_x) by means of the X^2 test:

$$X^2 = (F_v - f_v)^2/F_v + (F_x - f_x)^2/F_x$$

with one degree of freedom. A parallel analysis was done of the beginning of vireo songs relative to the periods of singing and silence of the flycatcher.

   Wasserman (1977) modified the method of Ficken *et al.* (1974) to compare the amount of time spent singing by four conspecific males. First, by measuring spectrograms of a large sample of songs (from 20 to 33 songs from each male), he calculated a mean value of the length (in seconds) of each male's song. Then, he multiplied this mean song length times the number of songs he heard each male sing during an observation period of 1185 minutes. He determined if neighboring sparrows avoid singing when conspecifics are singing.

   Let s equal the amount of time spent singing by any one of the four conspecifics, and s^o equal the periods of silence. The probability of one of the conspecifics being in song at any one instance is:

$$p(s) = s/(s + s^o)$$

The probability that none of the conspecifics is singing is:

$$p(s^o) = s^o/(s + s^o)$$

Suppose the particular individual being examined sings w total songs during the period of observation. If these songs are begun at random with respect to songs of conspecifics, the predicted number of songs from the individual sparrow (W_s) begun while a conspecific is singing is:

$$W_s = p(s)w$$

The predicted number of individual songs begun during periods of silence is:

$$W_s o = p(s^o)w$$

The expected values, W_s, and $W_s o$, are compared to the observed values, w_s and $w_s o$, using the chi-square test with one degree of freedom.

The observed to expected overlaps between individual pairs of birds were also compared. Here $W_{i,s}$ and $W_{i,s} o$ pertain to the expected number of songs sung by a bird under investigation while a neighboring individual, i, was singing, s, or silent, s^o.

(*Seiurus aurocapillus*) shortened their songs and sang less frequently and at more irregular intervals when song of another species was played through a loud speaker (Popp & Ficken, 1987). These examples tells us that the timing of song delivery by birds may be a means to overcome acoustic interference, just as it is in frogs. Masking is avoided by temporal separation as well as through the use of sound-frequency separation.

Acoustic masking is difficult to achieve experimentally in free-living animals. Bremond (1978) attempted acoustically to mask songs of the northern wren (*Troglodytes troglodytes*). He combined sympatric hetero-specific songs or altered wren songs with playback of the wren's territorial song. When a normal wren song is played back with a normal between-song interval of 8–10 seconds on a wild wren's territory, it takes only 1.48 songs before the defending wren has reached the speaker (about 14.8 seconds). Bremond was able to delay these appearances to 3.7 song playbacks when they were mixed with two other species' songs. However, he was unable to mask the wrens' song by mixing it with tapes of altered wren song. Thus the effect of a rich acoustic environment, one containing several other species singing, seems to hinder the wrens' ability to respond superquickly. Bremond suggests this is a 'psychological' hindrance, not one that is a result of acoustic masking. These results suggest that acoustic interference is based upon an alertness factor rather than simple acoustic masking alone – the animal hears the signal but takes longer to decide whether or not to attack. Perhaps locational cues are masked rather than perception of the species-specific signal.

An important addition to frequency-dependent attenuation is a consideration of what happens to the waveform and frequency content of a signal as it travels through the environment. When differential frequency attenuation within a signal occurs, the signal is said to be degraded relative to the acoustic form at its source (Wiley & Richards, 1978; Richards & Wiley, 1980). Other important sources of signal degradation occur through reverberations, reflections, and refractions. Degradation is a general term that encompasses any change (other than an even amplitude decrease over all component frequencies) in a signal from its emission to reception. Reverberations, amplitude fluctuations, and differential frequency attenuation cause changes in the signal (degradation) that are related to the distance the signal has traveled. Since the studies of Richards and Wiley (1980), several studies have sought to incorporate signal degradation into evolutionary considerations of signal design and long-distance communication, especially in birds. Just as the study of sense organs might open up new insight into the way animals view their world (Hopkins, 1983), the study of signal degradation has provided us with a new evolutionary perspective on birdsong function and evolution, called ranging (Morton, 1982; 1986; 1996a; McGregor, 1994).

3.8 Development of the ranging hypothesis for distance estimation

Why do songbirds learn songs, why does song repertoire size vary between and within species, why do individuals in some populations share all song types while in other populations they do not, and why is female singing so rare in temperate regions while common in tropical ones? Researching these questions, and others, relies on the ease with which song can be recorded and reproduced, manipulated, and stored. When used in playbacks, songs are unparalleled as natural stimuli (Hopp & Morton, 1997). As a consequence, birdsong research embraces nearly all levels of biological inquiry, from molecular to whole organism to population. There is great potential for birdsong research fully to integrate these levels and provide a holistic view of a complex behavior and its evolution.

From an assessment/management perspective, birdsong, and especially, ranging theory (Morton, 1982), provided an early warning that the informational perspective was inadequate. Ranging theory was the first new concept to underscore the importance of the assessor in the evolution of a form of communication and to support early efforts to change the guiding general concept of communication research away from an informational perspective (Owings & Hennessy, 1984).

3.8.1 A short history of explanations for song learning

Ranging theory has two lessons to impart. One is the perception of degradation by birds, the other concerns the evolutionary outcomes of this perceptual ability for A/M. One outcome is song learning, long a main focus of birdsong research (Marler & Hamilton, 1966). Song learning is found in some subgroups of three orders in the Class Aves (passerine or perching birds, hummingbirds, and, perhaps, a toucan in the woodpecker order Piciformes) (Nottebohm, 1975). Parrots (Psittaciformes) learn signals but for social use, not territorial defense. Song learning is rather uncommon in birds; most birds inherit their long-distance signals, but the phenomenon is well studied because song learning is suggested to share some attributes of human speech (e.g., Marler, 1975).

Here, an attempt is made to describe why, not how, birds learn songs. A common function of song, territorial defense, does not suggest why learning songs should be favored by natural selection. Song learning is typical of many families of birds termed oscine passerines. Their close relatives, the rest of the perching bird order (Passeriformes) do not learn songs, even though many sing in defense of territories just like songbirds. It is doubtful that these nonoscine passerines learn songs because their songs vary little geographically, and attempts to find evidence for learning have failed (see Kroodsma & Baylis, 1982; Kroodsma, 1984; 1985; 1989). Kroodsma and Konishi (1991) showed that a deafened nonoscine (eastern phoebe, *Sayornis phoebe*) still learned normal songs and lacked cell clusters like those in the forebrain song nuclei of songbirds.

Past attempts to explain why birds learn songs have stimulated research but have not resulted in a widely accepted synthesis. They are also few. Marler (1960) argued that learning enhances speciation because songs can change rapidly, producing species isolation. Nottebohm (1972) added that co-adapted gene complexes might be conserved by assortative mating among birds sharing the same song dialect – a song dialect is a population-restricted song type or types (see Marler & Tamura, 1962) – and that, in turn, dialects, might enhance speciation. Speciation, however, is abundant in nonoscine passerines, and dialects, while they are one result of the ability of oscines to learn songs, are not sufficiently widespread to be a cause of it. Furthermore, song learning does not appear to have contributed to the diversity of oscines by restricting gene flow (reviewed in Baptista & Trail, 1992). More recently, Nottebohm (1991; 1996) suggested that the dampening of sounds by the stapedius muscle of the inner ear, in order to protect the ear against overly loud self-generated sounds, could lead to auditorily guided vocal flexibility. Although his title suggests an analysis of the origin of vocal learning, Slater (1989) only speculates about current advantages that might maintain vocal learning. And, there is nothing new in the most recent book on birdsong (Catchpole & Slater, 1995). Hansen (1979) suggested that learning provides a filter so that only songs that transmit best are sung. Perhaps, if we find out why songbirds need such a filtering mechanism more than other birds, we can answer the question of the origin of song learning. Ranging theory may provide such a mechanism.

3.8.2 The development of ranging theory

A new perspective on the origin of song learning, called ranging theory (Morton, 1982; 1986; 1996a; McGregor, 1994) was stimulated by the finding that a bird, hearing the song of a conspecific, adjusted its responses dependent upon the amount of degradation in the song (Richards, 1981). Richards played a tape of 'near' and 'far' songs on the territories of wild Carolina wrens (*Thryothorus ludovicianus*), 25 m away from each bird. 'Near' and 'far' songs differed only in that 'far' songs contained reverberations and other sources of signal degradation that accrue with passage through 50 m of woods. The songs were equalized for amplitude. He obtained different responses. The birds sang after hearing the degraded songs, just as they respond to songs from neighbors a long distance from their territory. Instead of singing, they rushed to attack the speaker when they heard the undegraded 'near' songs. Richards concluded that Carolina wrens used the degradation in the song to range the distance to its source and to respond accordingly.

The informational perspective was used to interpret the results. In this case, it was suggested that songs conveyed information about the singer's distance to listeners so as to avoid costly fighting. Our A/M perspective emphasizes what communication accomplishes, suggesting that assessment may be more important. What are the responses desired by singers? Or, more explicitly, how are they attempting to manage the behavior of assessors? Do singers provide songs that are full of degradation and easy for listeners to range, or are these songs acoustically adapted to degrade little, making them sound as close as possible?

Study of the responses of wrens to *hearing* song was followed by a study of the *singer's* 'goals,' the other side of the coin, to differentiate between the two alternative explanations for the phenomenon of degradation assessment. Gish and Morton (1981) recorded Carolina wren songs from birds in deciduous forests in Maryland and 2200 km south in subtropical palmetto hummocks in Florida. The songs did not differ in frequency range or mean frequency so frequency-dependent attenuation could not contribute to degradation, regardless of the song's native habitat. Following Richards, they broadcast 50 songs and re-recorded them through 50 m of habitat in three sites, the same temperate deciduous forest, the same subtropical palmetto hummock the songs came from, and a third site where none of the songs was native. Gish and Morton compared the energy/time distribution in each undegraded song with its distribution after traveling through 50 m of habitat. This was accomplished by converting the recorded song into an amplitude/time trace using a sound-level recorder. Each trace consists of a series of peaks, each peak representing an element in a song. They then compared the distribution of these peaks before and after the songs were re-recorded and converted to an index of how much change occurred.

This result conforms to the A/M prediction. Rather than being designed to inform listening males of their distance from the singer, songs hide this as much as possible. From the assessing wren's standpoint, hearing a conspecific song is like listening to the whistle of an incoming mortar round. It is crucial to become a good estimator of distance! A good listener is one that uses the least amount of energy to defend its territory. Judging distance to possible invaders is a valuable tool in this time/energy conservation, and one that the signaler should make difficult, thereby causing the listener to invest time and energy to investigate. The results of this study showed that Carolina wren songs are physically structured to retain their source characteristics during propagation. The set of Florida songs had lower mean change index values in the Florida test sites (0.208) than in the deciduous forest site (0.220), while the deciduous forest songs had lower index values in their native site (0.198) than when tested in Florida (0.236). Change index values for the 'neutral' sites showed no difference in degradation in the two song groups. Songs native to the test area degraded less than songs foreign to the area. Clearly, Carolina wrens' songs are adapted to degrade as little as possible, substantiating Hansen's (1979) idea that song learning could provide a means to adapt song to habitat acoustic conditions. Finally, in a laboratory study, we showed that song-learning wrens use song degradation in their choice of songs to learn (Morton, Gish & Voort, 1986).

We have two sets of data on song, one on the responses to song showing that degradation is a factor (Richards, 1981), and one showing that the songs do not degrade predictably to the listener (Gish & Morton, 1981). This suggests that listeners might use general aspects of degradation to estimate distance. Further evidence proved this was not the case in Carolina wrens (Shy & Morton, 1986a) and other species tested (Sorjonen, 1983; McGregor, Krebs & Ratcliffe, 1983; McGregor & Falls, 1984; McGregor & Krebs, 1984; Morton & Young, 1986; Shy & Morton, 1986b).

These playback studies showed that the perception of distance using degradation in signals is based upon whether or not the listener has the perceived signal in its memory. General features of degradation do not account for the field data reported. Most of the studies used playbacks of songs to show that degradation became an effective contributor to responses only if the focal individual had the song type it heard in its memory. Generally, the playbacks were done in the center of a focal bird's territory, and the researchers compared responses to undegraded and artificially degraded songs played from the same position. Only one study used playbacks from within and without territories, thus providing natural degradation for the birds to assess (Shy & Morton, 1986a). However, all studies have shown that the bird must have the signal in its memory before it is able to assess degradation and make responses appropriate to the apparent distance from which the signal originated. Estimating distance to singer by comparing degradation in perceived songs to an undegraded version of it in memory is called *ranging* (Morton, 1982). How this might work is discussed below.

3.9 Ranging, the arms race between manager and assessor, and the learning of long-distance signals in birds

We know that ranging evolved before songbirds evolved the ability to learn songs. Ranging is found in perching birds (passerines) whose songs develop without the need for the bird to hear them (Morton & Derrickson, 1996). This means that ranging did not lead, inexorably, to song learning, as found in songbirds. What is necessary for song learning to be favored is an evolutionary arms race between song function and performance for managers and ranging ability in assessors (Dawkins & Krebs, 1979; see Chapter 1). Ranging by assessors probably started the arms race, an example of the importance of assessment that is highlighted throughout this book.

We still do not know how, neurophysiologically, distance perception through ranging is accomplished. The process may be similar to bat echolocation but, instead of sending out a known signal and using arrival time, the bird assesses the amount of degradation (approximating its distance from the source) in the incoming signal with its undegraded stored version. More recent work suggests songbirds may integrate received songs over longer time courses (Margoliash & Fortune, 1992). The bird might use motor inputs to the neural tissue, activating its memory of what its own song sounds like when produced. Williams and Nottebohm (1985) suggest that syringeal (the syrinx is the sound-producing organ in birds) hypoglossal motor neurons respond selectively to natural song elements heard by the zebra finch (*Poephila*). Thus, listening or singing finches have similar neural circuitry and it should be possible for a listener to convert a song heard into the motor commands necessary to reproduce the same sounds. If this is the case for ranging, only stored songs will allow an assessment of degradation. Margoliash (1983) supported this notion. The auditory response properties of units in a telencephalic nucleus in white-crowned sparrows (*Zonotrichia leucophrys*) exhibited considerable selectivity for the individual's own song. Furthermore, song-specific units in wild-caught birds showed intradialect selectivity (a birdsong dialect is a population-restricted song type or types). It thus seems reasonable that the songs a bird hears are compared to song(s) stored in a portion of the brain termed the HVc and that these autogenous (self) songs serve as the reference component in ranging (Margoliash & Konishi, 1985; Margoliash, 1986; Margoliash & Fortune, 1992; Margoliash *et al.*, 1994). The Gambel's white-crowned sparrow (*Z. l. gambelii*) offers an opportunity to test the generality of autogenous song sensitivity because, in this population, dialects are not found and self songs are individualistic (Austen & Handford, 1991). Perhaps response to autogenous songs is lessened here?

Since the perception of degradation involves precise time assessment, it is probably no coincidence that birds are superior to most mammals only in this area of peripheral hearing ability. Time-interval assessment takes place peripherally, e.g., in the ear, rather than in the central nervous system (Konishi, 1969). Budgerigars (*Melopsittacus undulatus*), for example, can resolve sounds separated by as little as 1–2 ms. Humans, by contrast, lose sensitivity to sounds happening faster than 5–6 ms apart (Dooling, 1982).

To summarize, birds may be quite good at using the degraded structure of an incoming signal to approximate the distance it has traveled. What does the perceptual mechanism involved in ranging mean to the evolution of long-distance signals? This very general question suggests many avenues for future research and, while we know little about the mechanisms of ranging, the logic of natural selection allows some further predictions of ranging theory. Whether these predictions turn out to be correct depends upon the development and testing of specific hypotheses.

3.9.1 A conceptual framework for the evolutionary predictions of ranging theory

It is not surprising that birds can range song distance for it provides them with a means to avoid wasting energy. Indeed, ranging may be a general ability of all birds, not just songbirds (Morton, 1986). Assessors should adjust their responses to singing to the threat the singer poses to their territorial or reproductive interests and they should not expend more time and energy than is necessary to defend these interests. One possibility, therefore, is that ranging evolved because it served this interest for assessors: ignoring songs is important to their energetic balance when it allows uninterrupted foraging, mate guarding, etc., without incurring costs. Ranging evolved because it served this function for assessors, but such ranging did not favor the evolution of song learning by managers.

It is likely that long-distance signals evolved to control space by making an individual's presence *detectable* throughout that space. Why didn't evolution simply stop there? After all, a detectable 'keep out!' can function to reduce the cost of territorial defense. But ranging theory predicts that advertizing an individual's presence has long disappeared as the most important factor in *small* birds. Birdsong is too structurally complex to be explained fully by either detectability or species-isolating mechanism arguments.

'Detectability' refers to the maximum distance from the singer at which its song can be differentiated from background noise and recognized as to species. Detectability is not sufficient to explain birdsong evolution because selection pressures are generated if a bird responds to the song of another that is not a threat to that bird's territorial space, or parentage, because *time and energy are wasted*. Selection will favor individuals with means to avoid wasting energy in this manner. Since degradation provides an estimation of distance, selection should favor a listener's use of degradation to determine if energy would be wasted if it responded to a song. This means that the distance from a singer at which a signal can be barely detectable becomes a minor source of selection on the structure of songs. The distance through which a song functions effectively to further the goals of the singer may be much shorter than that song's distance of detectability because of ranging by assessors.

Earlier, we mentioned that singers and listeners have different 'goals' that led to an evolutionary arms race between the roles of singer and listener. What constitutes the arms race? When assessors range, they are not disrupted by singers and conserve energy but this, in turn, reduces the distance to which songs effectively threaten, forcing singers to move closer to rivals in order to defend territory. Singers evolved counter-adaptations to ranging.

3.9.2 *Counteradaptations to ranging*

Singing reflects these counteradaptations, and some characteristics that we take for granted, such as the complexity and beauty of birdsong, are due to them. If assessors use degradation assessment mechanisms (DAMs), a manager should use a song that 'sounds close' to the recipient even though the singer may not be close. Such a song would be structurally adapted to degrade as little as possible in the singer's local habitat and have a high source amplitude. Then, a song will propagate as far as possible and still remain relatively undegraded. As we have seen, Carolina wren songs are adapted in just this way. Evidence from field and laboratory studies of song degradation support the prediction that 'sounds close' (Hansen, 1979; Gish & Morton, 1981; Morton *et al.*, 1986; Shy & Morton, 1986a) is a function of birdsong, in some oscines, that results in acoustic adaptations to habitats.

The arms race produced a second way to thwart listeners' DAMs: sing songs not in the listener's memory. By using songs differing from their memorized songs, assessors cannot range them effectively. One result of this mechanism is the development of song repertoires. The function of large repertoires of songs and high internal song complexity (e.g., songs composed of many, temporally ordered, syllable types) can be understood in relation to the listener's need to use its memorized songs to range.

3.9.3 A graphical portrayal of A/M conflict in birdsong

'Sounds close' describes a listener's assessment of a song, whereas 'song distance' is a measure of distance from the singer to the point at which the song will not be perceived as close. Assessors will ignore them at that point.

'Sounds close' and 'song distance' should be viewed as testable theoretical concepts derived from perceptual predictions of ranging theory. Their interrelationships are shown diagrammatically in Figure 3.13 where:

d = distance from singer wherein the song is detectable
a = 'song distance' – distance from singer wherein 'sounds close' can function (i.e., listeners will not ignore the signal)
$m = d - a$

a and d are dynamic, changing among acoustically different environments and as variable meteorological conditions affect sound transmission at a given site. However, the value of a/d, on average, may differ consistently between environments. For example, a is generally greater within closed canopied forests than in grasslands. Another way to say this is as $a \longrightarrow d$ ($m \longrightarrow 0$), sound frequency-dependent propagation becomes more important as a source of selection on signal structure because the use of least attenuating frequencies is one way to increase d without increasing degradation. As m increases, sound degradation increases in importance as a source of selection on signal structure. This is supported by empirical data: intraspecific comparisons show that birds living in relatively acoustically homogeneous environments have songs of narrower frequency range than conspecifics living in acoustically heterogeneous environments. Species living in forests have songs with a narrow frequency range matching 'sound windows' with lower attenuation, while species living in open habitats (acoustically heterogeneous) do not. These empirical observations support the suggestion that the value of m and its variance are an important predictor of long-distance signal structure.

We focus again on the singer/listener dichotomy in Figure 3.13B and C where t, a territorial boundary, is added. For an 'ideal' singer, $a = d$ and $m = 0$ (Fig. 3.13B). For such a singer, *song distance* would be equivalent to the maximum distance at which its song is detectable and no listener within the radius of detectability would ignore the signal. For an 'ideal' listener, $m \longrightarrow d$ (Fig. 3.13C). Selection should favor both the acoustic structure of signals that increase a and the timing of singing to match

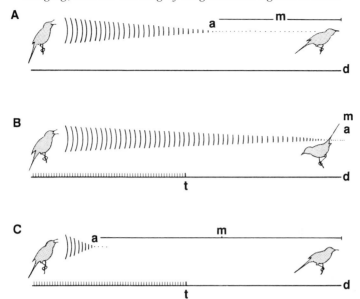

Fig. 3.13. The difference between detectability and effective song distance in ranging theory. (A) The relationship between the distance a song is functionally effective (a), the distance to which it propagates and is detectable (d), and m, the difference between d and a as perceived by a listener. (B) The singer would prefer that its song distance be the same as its detectability (a = d, m = o), especially in reference to its territorial boundary (t). (C) The way in which a listener would prefer to perceive the song: to ignore it as long as the singer is not encroaching. This 'battle' of self-interest illustrates a classic case of assessment influence on management efforts and contributed to the learning of songs by managers in small songbird species (Morton, 1996a).

meteorological conditions conducive to increasing *a*. By adding a territorial boundary, *t*, we can now state that the singer's goal is to increase its *song distance* beyond *t* (say to *t = d*), whereas the listener's goal, by using DAMs, is to reduce the distance from *t* to *d* to 0 (in other words, to reduce *song distance*, *a*). Any success singers have in increasing *a* beyond *t* will cause energy wastage in listeners; in short, their ongoing behaviors will be disrupted to the singer's advantage.

3.9.4 Disruption and threatening functions of birdsong

Disruption to the listener is predicted by ranging theory to occur whenever *a > t*. This is an important prediction for two reasons. First, it may represent the asymmetry needed to favor the evolution of DAMs, as

mentioned above. Second, it suggests that ecological variables are important in the evolution of long-distance signaling, because birds on low-
quality territories will not be able to afford time to sing relative to those
on food-rich territories.

Two main functions for territorial singing, threat and disruption, differ
in the following way: threat says only 'I am here' or, 'I am here and the
singer is actually close,' whereas disrupt says 'I am here and I am close'
when, in actuality, the singer is not close.

Song distance took on additional importance when assessors evolved
ranging because singers could now threaten rivals without the energy cost
entailed in locomotion. When border disputes occur, rivals can provide
accurate distance cues through ranging. Songs function to threaten more
effectively if the singer uses songs found in the rival's memory because
these can be ranged better than if nonshared songs are used. This is just
what we observe when two birds interact over boundaries: they switch
from song types used before the interaction to types that 'match.' This is
commonly termed matched countersinging in the literature on birdsong.
In species having repertoires of several song types, matched countersinging involves a nonrandom choice from the several types in memory to
match a rival's song. This was viewed as a form of escalation of threat but
the precise reason for matching was vague (Krebs, Ashcroft & Orsdol,
1981). Informational explanations of matched countersinging were not
efficacious, but our A/M approach easily encompasses the importance of
assessment to the function of matched countersinging.

The function of ranging in matched countersinging is closely related to
the evolution of song dialects in songbirds. Dialects are defined as 'a
consistent difference in the predominant song type between one population and another of the same species' (Marler & Tamura, 1962). Dialect
systems differ from geographic variation in that a sudden change in song
type(s) occurs often with no geographic separation between the adjacent
dialects. A critical feature is the absence of unshared songs. Ranging
theory predicts that dialects will be adaptive where and when threat,
i.e., 'honest' signaling of distance to listeners, is the primary function
of singing behavior. Dialects might be called permanent matched countersinging. They occur in areas of mild climate where neighborhoods are
stable. That is, turnover of territorial neighbors is so low that territorial
boundaries are known by rivals and may last for many years. Because of
known boundaries, rivals essentially use matched countersinging to
threaten one another. Consequently, dialect species respond most
strongly to within-dialect songs that they are able to range rather than

to unfamiliar conspecific songs they are less able to range. The boundaries are defended and the advertizement of territorial quality is of lesser importance. Switching territories, rather than enlarging them and changing boundaries, like repertoire singing species do, may be the norm for dialect species. This prediction has considerable support (Gompertz, 1961; Lemon, 1967; Bertram, 1970; Milligan & Verner, 1971; Harris & Lemon, 1974; other references in Kroodsma, 1976; but see Craig & Jenkins, 1982; Ratcliffe & Grant, 1985).

The geographic distribution of dialects supports these predictions. Bitterbaum and Baptista (1979) illustrate several examples of intraspecific dialect occurrence or absence. In all cases, eastern or northern North American populations do not have dialects whereas populations in warmer climates have dialects. In some cases, this correlation entails migratory versus nonmigratory populations, with nonmigratory populations, probably with more stable neighborhoods, having dialects. An example is found in the white-crowned sparrow (*Zonotrichia leucophrys*), whose well-studied song dialects are found in populations inhabiting the warm climate of California (Marler & Tamura, 1962; Baker, 1974; Baptista, 1975) but not in migratory populations of temperate eastern North America (Austen & Handford, 1991).

3.9.5 *An ecological foundation of communication*

Ranging theory predicts that the threat and disrupt functions of long-distance signals have an evolutionary origin in ecology, not communication. Disrupting territorial rivals may accentuate any differences, however small or temporary, in food resources on territories (Fig. 3.14). High song output advertizes lots of food when singing and foraging for food are mutually exclusive. Time spent singing provides assessors with a direct measure of the excess time available for feeding nestlings. Each morning birds are at their lowest energy period of the diurnal cycle (Kacelnik & Krebs, 1982). This would be the time for a bird to 'flaunt the quality' of its territory for attracting females for pairing and extra-pair copulations as well as mate guarding (reviewed in Cuthill & Macdonald, 1990).

The idea that birds on food-rich territories should sing more than individuals on poorer territories has been tested by providing food artificially. Song competition among birds is basically an energetic contest. The bottom line is 'if you've got it, flaunt it', 'it' being a more food-rich territory. Reid (1987) studied this in savannah sparrows (*Passerculus*

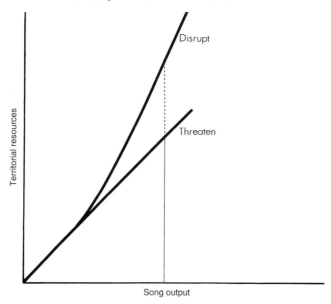

Fig. 3.14. Ranging can increase the effectiveness of singing when singers use unrangeable songs or those that have less reverberation (see text). This model shows the relationship between how long and how much a bird sings (song output) and territorial resource levels when song is used only to defend territorial boundaries (threaten), a linear relationship, versus when song can disrupt the activities of listeners because they are unable to range its distance (exponential curve). Females choosing social mates or extra-pair mates are predicted to prefer males with greater values along the resource axis for a similar song output. The disparity between the two curves (dashed line) ranks males on their song output as an index of male fitness and territory resources, as does song output alone.

sandwichensis) on Sable Island, Nova Scotia. She found that song cost was reflected in the effect of energetic stress on song rate. After cold nights, males weighed less and sang less. Providing extra food on some territories increased song rate and decreased foraging. Additional research using food supplementation supported the idea that song output is maximized so as to advertize food availability, with males 'overexerting' themselves when singing until a mate is acquired (Reid, 1987; Gottlander, 1987; Radesater *et al.*, 1987; Strain & Mumme, 1988; Alatalo, Glynn & Lundberg, 1990; Arvidsson & Neergaard, 1991; Hoi-Leitner, Nechtelberger & Hoi, 1995). Furthermore, in the permanently paired Carolina wren, Strain & Mumme (1988) confirmed that song output is limited by food availability.

Females prefer to mate with territorial males that spend the most time in song, when singing and foraging cannot occur together (in species with short-term pair bonds formed only for breeding, not those with year-long pair bonds). The intervals between songs (silent periods) are so regular that these may allow females to compare song output among males (Beletsky, 1989). Thus, because singing is costly in time, song output becomes a reliable indicator of territory quality. Song function, therefore, has an ecological base.

3.9.6 Ranging and other hypotheses on the evolution of long-distance signals in birds

Historically, discussions of the evolution of avian song have treated diverse aspects of singing behavior, for example song dialects and repertoires, as separate issues. There was, seemingly, no concept of sufficient generality to link them as different outcomes of selection based upon a common perceptual mechanism. The following is a synopsis of older hypotheses to explain repertoires (Dawson, 1982) and then dialects (Payne, 1981):

Repertoires enhance recognition. Emlen (1971) suggested that factors promoting the distinctiveness of an individual's songs may allow others to more easily identify that individual. This is often called the neighbor–stranger hypothesis and will be discussed further below.

Repertoires evolved in response to intersexual selection. Once good-quality males start being more successful by using more complicated songs, then the runaway process of intersexual selection (Fisher, 1930) leading to extreme elaboration could be set in motion.

Repertoires increase success in territorial competition. This idea suggests that individuals with large repertoires may be more successful in territorial competition.

Repertoires allow the use of different songs in different contexts. Two levels of contexts have been implied under this idea: the use of different songs in the center of the territory versus border clashes (Lein, 1978) or a male's attentiveness to nest defense (Smith, Pawlukiewicz & Smith, 1978), and whether a male is paired or not (Morse, 1967).

Repertoires prevent habituation. This was originally presented as the antimonotony principle (Hartshorne, 1973) – *the beau geste hypothesis.* The antihabituation idea led to the suggestion that song repertoires in some species have evolved in the context of density assessment. Krebs (1977) felt that multiple song types repeled new males from settling by providing them with an elevated perception of the density of males already defending territories.

The following is a series of hypotheses to explain the origin of dialects.

The acoustic camouflage hypothesis. The deceptive mimicry hypothesis of Payne (1981; 1982) predicts the sharing of song types in neighboring males. He showed greater reproductive success in yearling males that match songs of older neighbors. He suggested that the increased reproductive success involved competitive mimicry and deceit of other males through mistaken identity based on the older territorial model's song. Craig and Jenkins (1982) suggested that new birds, by learning local songs from old birds, produce selection pressure on old birds to sing more complex, difficult-to-learn songs. In this way, the asymmetry between old and new birds can be maintained to the advantage of older birds.

The neighbor–stranger hypothesis. The most widely accepted idea concerning a song function is based upon the responses of resident birds to song playbacks. The strength of response has traditionally been used to tease out 'species recognition' properties of birdsong structures. Unfortunately, the song structures used were rarely specified exactly and the degradation in the test songs was not standardized. However, researchers have consistently found that the response of resident birds is greater to the playback of a stranger's song (noncontiguous neighbor) than it is to the playback of a firmly established territorial neighbor's song. The lowered response is thought to conserve energy amongst males with well-established territories by reducing strife (Weeden & Falls, 1959). This is also called the 'dear enemy' hypothesis (Fisher, 1954). 'Neighbour' or 'stranger' classifications of signals, however, do not allow a rigorous discussion of the effects of signal structure on responses even though such discriminations may exist along with ranging (Falls, 1982).

Differing song structures are epiphenomena of song learning, 'cultural' phenomena with no biological basis. This idea is the only one that is incompatible with RT: the assumption that signal sharing represents a cultural epiphenomenon as a byproduct of song learning. The idea that specific song types represent cultural adaptations and that their physical structure is therefore arbitrary (Mundinger, 1982; Payne, 1996; Lynch, 1996) must be re-evaluated in light of the specific functions for song structures proposed in ranging theory. The biological significance of song type(s) held in common is to permit a singing individual to threaten others in a very specific sense. The singer is not simply proclaiming its presence, which could be accomplished through any species-specific signal, as was discussed at length above. Using energy efficiently to manage conspecifics via ranging is the reason that signal structures are not arbitrary.

3.9.7 Ranging theory and song learning in birds

The ranging theory suggests that the evolutionary response of managers to ranging was a key to the evolution of vocal learning. This response took two main forms: the acoustic adaptation of songs to avoid degradation in local areas, to increase *song distance*, and the evolution of song complexity and repertoires to lower ranging accuracy. The close interactions between management and assessment resulted in song learning because songs that are developmentally inflexible could not be so precisely adapted.

There is still controversy, however, about whether listeners can learn to range songs they hear from neighbors but do not sing themselves. McGregor (1992) felt that his study of great tit (*Parus major*) responses to song degradation and song 'familiarity' (McGregor & Krebs, 1984) proved that neighbors could range each other's songs even if they do not share them. They compared responses to three categories of songs: those found only in the test male (OWN), those found in both the test male and one of his neighbors (OWN + NEIGHBOR), and those found only in neighbors (NEIGHBORS). Only two of eight measures showed significant heterogeneity, and they concluded that their three categories of familiar song had little effect on degradation discrimination. They also concluded that songs of neighbors can be ranged even if the test male does not sing them (McGregor, 1991). But, to see if birds respond similarly to songs that they share with their neighbors, and to songs sung by neighbors but not by themselves, one should compare only the categories OWN + NEIGHBORS and NEIGHBORS, not their responses to OWN songs too, which are obviously in the bird's memory.

When this is done, one finds for undegraded songs that the birds respond significantly more strongly to OWN + NEIGHBOR than to NEIGHBOR songs, and that the discrimination between these two categories is very diminished in response to degraded songs (Eyal Shy, unpublished manuscript). In other words, birds *can* sing unrangeable

(Zahavi, 1977; 1987), and little or no selection pressure on singers to use countermeasures to ranging exists. Indeed, selection has favored singers who *use* the ranging (assessment) ability of listeners against them: they threaten by producing songs rangeable to everyone in the population (Morton, 1986). These songbirds use song in the same manner that non-learning nonoscine passerines do. It is not accidental that dialect-song-birds share the same stable or tropical climates which host most of the nonlearning nonoscine species (Morton, 1986).

The ranging hypothesis offers an explanation for many, often dispa-rate, functions of birdsong as well as the general trends already discussed. Playback experiments have supported several earlier predictions (Morton, 1986). For example, in addition to oscines, which are generally small species, evolutionary arms races should also produce song learning in other small-bodied, nonoscine taxa. This was confirmed for a hum-mingbird (*Calypte anna*) (Mirsky, 1976; Baptista & Schuchmann, 1990), which is energy limited (Stiles, 1971). The presence of song dialects in lekking hermit hummingbirds (e.g., *Phaethornis longuemareus, P. super-ciliosus*) suggests that song learning may be widespread in hummingbirds (Snow, 1968). Ranging has recently been shown in a nonoscine passerine, the dusky antbird (*Cercomacra tyrannina*), supporting the prediction that ranging evolved before song learning did in the passerines (Morton & Derrickson, 1996).

4

Mechanisms and proximate processes of vocal communication

This chapter is not divided as Chapter 2 is, into separate sections on assessment and management. Instead, it considers the integrated proximate functioning of these two processes, discussing the contributions of perception, cognition, motivation, and emotion to the interacting operation of assessment and management both within and between individuals. These multileveled interactions are most clearly highlighted in the last section of this chapter, in which the development of communicative abilities is explored. The adoption of an A/M approach has provided a framework for dealing with the dynamics of social interaction.

The material here may seem a bit off the point for those readers expecting a treatment dealing exclusively with vocal communication. Time is spent exploring broader issues, of how perception, cognition, motivation, and emotion work in general, not just during communication, and of how interactions between individuals not only result from management and assessment, but also influence those two processes. This latter point, that the dynamics of interaction among individuals is the arena of communication, is convergent with the same consistent theme in the writings of John Smith (1997), even though its source lies in a very different, noninformational starting point from Smith's. We share with Smith this message: if we are to continue progressing in our understanding of communication, we must return systematic descriptive research on the broader contexts of communicative behavior to its former high-priority position. With this theme, we ally ourselves once more with Tinbergen (1963), whose seminal paper on the aims and methods of ethology featured not only sections on his four questions, but also an initial section entitled 'Observation and Description.'

4.1 Perception

We all know that the fundamental function of perceptual systems is to support the actual pickup of cues. Without auditory systems, animals could not use acoustic cues because they could not detect sound. But the role of auditory systems is more subtle than that. As discussed briefly in Chapter 2, perceptual systems are designed, through both selection and proximate effects of interactions with the environment, to be selectively attuned to those features of the environment most important to the regulatory problems faced by individuals. So, we can use the response properties of auditory systems as clues to which features of vocalizations are the salient ones. Conversely, we can use the features of vocalizations as guides in our exploration of the tuning properties of the auditory system. Attunement of hearing to vocal systems has been discovered at multiple levels in vertebrate auditory systems, from the ear to central brain structures, and at multiple levels of complexity in sound structure.

The *audiogram* provides an example of tuning in part from a peripheral structure in the auditory system, the ear, and a simple feature of sound, its frequency. The audiogram is a measure of the auditory sensitivity of an individual to each of a range of sound frequencies. Audiograms typically reveal frequency tuning, that is, sensitivity peaks at one or more sound frequencies (Fig. 4.1). Correspondences have been discovered between these sensitivity peaks and the dominant frequencies of such important sounds as the vocalizations of conspecifics. The responses of female green treefrogs to male courtship calls illustrate selective responsivity founded in part on the audiogram (Gerhardt, 1987). The inner ear of frogs has two separate receptor organs: the amphibian papilla, which is tuned to sound frequencies of 0.7–1.2 kHz in green treefrogs, and the basilar papilla, which is tuned to 3.0–3.6 kHz in green treefrogs. These peaks correspond approximately to the spectral peaks in the calls of average-size males. Females show a preference for those sounds that most stimulate their auditory systems, and the tuning of the inner-ear organs appears to play a role in mediating the selectivity of females' responses to male calls. The results of playback studies have been complex; nevertheless, when females have exhibited a frequency-related response, they have preferentially approached sounds with frequency characteristics that match the tuning of their ears. Such selectivity in the auditory system may be one mechanism whereby females avoid sexual approaches to males of the wrong species.

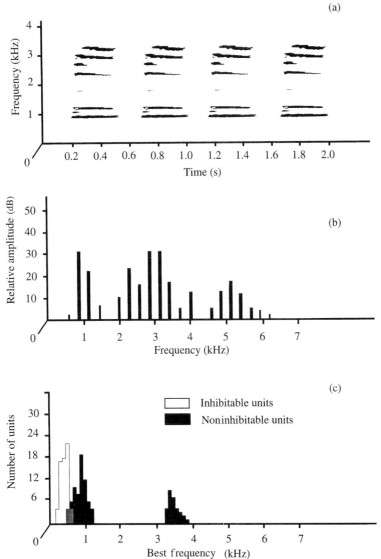

Fig. 4.1. An example of a match between the spectral structure of the calls of male green tree frogs and the spectral sensitivity of the auditory system of the target of these calls. (a) Four consecutive calls by a male. Note the concentration of the two clusters of call components around 1 and 3 kHz in the frequency spectrum. (b) This graph confirms the concentration of acoustic energy in two clusters around 1 and 3 kHz, respectively (as well as a third cluster around 5 kHz). (c) The selectiveness responsiveness of the green tree frog auditory system. The (noninhibitable) auditory neurons are most responsive to sound frequencies in the vicinity of 1 and 3 kHz. (Drawing (a) courtesy of Carl Gerhardt; (b) and (c) courtesy of Robert Capranica.)

Auditory systems may also be attuned to more complex patterns of acoustic stimulation. For example, male swamp sparrows and song sparrows differ in the complexity of the songs that they sing, and exhibit corresponding contrasts in their innate bases for distinguishing each other's songs (Marler & Peters, 1989). The songs of these species are structured hierarchically, consisting of *notes*, which are organized into *syllables*, that are assembled into *phrases*, which in turn constitute the *song* (Fig. 4.2). As can be seen from Figure 4.2, swamp sparrows sing relatively simple songs, consisting of a single phrase in which the same syllable is repeated. In contrast, song sparrows typically sing four-phrase songs, starting with a *trill*, a type of phrase that is alternated with *note-complex* phrases. As discussed in Chapters 1 and 3, all songbirds studied to date need to hear conspecific song in order to develop normal song themselves. As young swamp sparrows undergo the early perceptual-learning part of this process, they distinguish swamp sparrow song from that of song sparrows on the basis of syllable structure. Song sparrows at the same developmental stage, on the other hand, recognize conspecific song on the basis of the segmentation of the song into phrases. If swamp sparrow syllables are embedded in such songs, for example, they will be learned by the developing song sparrows.

We must not forget, of course, that auditory systems are used in many other contexts besides vocal communication, and that these other acoustic contexts also may have shaped the properties of these auditory systems through development, and natural and sexual selection. So, there are limits to the extent to which we can expect to find tight links between the response properties of auditory systems of a species and the structure of their vocal signals. For this reason, studies of an auditory system should explore its responsiveness to a broad area of acoustic space. Such research has provided not only a more complete understanding of perceptual systems, but also a bridge between the study of proximate and ultimate processes (Ryan, 1994).

The phenomenon of 'supernormal' stimuli illustrates how such proximate–ultimate bridging can be accomplished. Ethologists discovered long ago that stimulus objects that extend well above the normal range of variation in quantitative dimensions such as size can be unusually effective in evoking reactions (Tinbergen, 1951). Perhaps the most commonly encountered cases of supernormal stimuli in the literature involve preferences by birds for incubating very large eggs. Herring gulls, for example, look incongruous attempting to incubate a model of an egg as large as 20 times normal size (nearly their own body size!) and ignoring

Song sparrow song structure: patterning of phrases
(4 phrases)

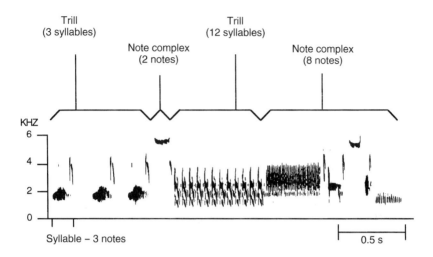

Swamp sparrow song structure: patterning of phrases
(1 phrase)

Trill (11 syllables)

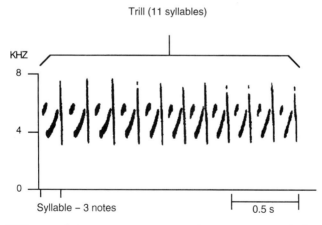

Fig. 4.2. Differences between song sparrows and swamp sparrows in the complexity of their songs. Song sparrows exhibit four organizational levels in their songs: notes, syllables (a set of notes), phrases (a set of syllables), and the song (a set of phrases). The songs of swamp sparrows involve only two levels, one phrase consisting of a string of similar notes. (Courtesy of Peter Marler.)

the model of normal size right beside it (Tinbergen, 1965, pp. 78–9). It is not clear why perceptual systems are so commonly hyperresponsive to hyperstimuli, but this trait, whatever its origins, may be a source of selection on signals, shaping natural features of animals into exaggerated versions of themselves through a process of selection for signal value. For example, a number of species of cuckoos exploit other avian species by laying their eggs in the nests of those other species, thereby inducing them to care for cuckoo offspring. The eggs of these brood-parasitic cuckoos are accepted by their hosts in part because they have evolved to mimic the colour patterns of the host eggs. But, the cuckoo eggs are also usually larger than the host eggs, perhaps reflecting adaptation to act as super-normal evokers of incubation behavior, providing the parasitic eggs with an edge over the host's own eggs (Wickler, 1968). Supernormal respon-siveness to exaggerated stimuli also appears common in the acoustic domain. For example, more than 150 studies have demonstrated female preferences for extreme values of optical and acoustic male traits impor-tant as cues in mate choice (Ryan & Keddy, 1992). Where such prefer-ences exist, the overwhelming majority are for values above rather than below the mean. Females prefer more – more intense, more complex, and longer calls and songs, as well as vocalizations delivered at greater repeti-tion rates. Perhaps these properties of female perceptual systems have been sources of sexual selection on the calls of males, favoring the evolu-tion of extreme signals for use in courtship. It is only through detailed studies of such mechanisms that we can come to a more complete under-standing of why evolutionary changes take the directions they do.

An A/M approach directs our attention to a set of cues that is more complex yet – feedback, or relations between output and input. We have already encountered the phenomenon of feedback. The ground squirrel in the Prologue, for example, assessed the level of risk she faced from the intruding rattlesnake by inducing it to rattle. She used the structure of the rattling sound to judge the snake's size and body temperature, two major determinants of the snake's threat to squirrels (see Fig. 2.11).

Such relational cues have traditionally not been available to the sub-jects of playback studies, the most popular type of experimental proce-dure in research on vocal communication (see Box 1.1; McGregor, 1992). As already mentioned, in a typical playback study, an audio tape is made of one or more vocalizations, and this tape is played to the subject irre-spective of how it responds. As noted in Chapter 1, portable computers have made it possible to use interactive playback procedures, in which the choice of playback sounds varies, depending on how the subject is responding. If relational cues are important, this should be revealed in different effects of traditional and interactive playback methods.

Work with European blackbirds supports the above prediction (Dabelsteen & Pedersen, 1990). Males of this species produce three structurally different types of songs – low intensity (LI), high intensity (HI), and strangled song (SS) – which are associated with low, medium and high probabilities of attack, respectively, on the part of the singer. When two males engage in a singing duel, they often adjust their singing to the behavior of each other in quite precise ways. Do such contingent adjustments make a difference? They can. The responses of males to song playbacks were compared under three conditions: (1) interactive shifts in song type, in which playbacks began with LI, were switched to HI if the target bird responded, and were escalated to SS if the subject approached the playback speaker to within 10 m; (2) noninteractive changes in song type irrespective of subject response; and (3) no change in played-back song type. Under all conditions, LI released the weakest aggressive responses, and HI and SS the strongest. But, HI and SS were distinguished only under the interactive conditions. So, the occurrence of feedback attunes interactants more precisely to the details of the other's behavior, perhaps making management more effective. It is in such interactive forms of perception that we see most clearly the links between assessment and management.

The effects of feedback on vocal communication have been studied systematically, most often in developmental time frames (e.g., Clayton, 1994). Nevertheless, many playback studies of animal vocalizations demonstrate the reality of immediate social feedback to vocalizing animals (e.g., McGregor, 1992); playbacks are used in studies of vocal communication *because* they have immediate social consequences. For example, male white-crowned sparrows respond to the playback of conspecific song by approaching the speaker, flying aggressively at it, and countersinging, even when they are only a few months old (DeWolfe, Baptista & Petrinovich, 1989). And, singing by these youngsters has the potential to evoke attacks by adult males. The developmental effects of such feedback are discussed later in this chapter.

Do animals adjust their communicative behavior to such consequences in more immediate time frames? Natural observations certainly make it appear so. The responses of other males to singing by young white-crowned sparrows typically is associated with a counterresponse by the young singer. But, do systematic experiments support this evidence? The data are not plentiful, but they indicate that immediate adjustments do occur. Among songbirds, chaffinches (Stevenson, 1967) and zebra finches (ten Cate, 1991; Adret, 1993) treat the playback of conspecific song as rewarding; in laboratory experiments, they perform activities such as key pecking and landing on a perch more frequently when these activities are followed by song playback.

The idea of feedback sensitivity suggests a literature to consult for guidance in exploring the features of perceptual mechanisms in vocal communication; auditory mechanisms of echolocation are of necessity sensitive to feedback (Neuweiler, 1990). Indeed, from an A/M perspective, echolocation and vocal communication bear some striking resemblances (West & King, 1990; Tyack, 1997). In both cases a major source of assessment cues arises from attention to feedback from emitted sounds. In echolocation, the feedback is from the echo of the emitted sound. In vocal communication, the feedback is from the reactions of other individuals to the vocalization.

Can the literature on echolocation suggest design features to seek in vocal communication perceptual systems? Perhaps. For example, one of the problems faced in echolocation is interference between the auditory effects of the high-intensity emitted vocalization and the low-intensity echo; the effects of the emitted sound have the potential to mask detection of the echo (Neuweiler, 1990). The results of electrophysiological studies of auditory neurons in horseshoe bats are consistent with this expectation: neuronal response to a tone can be suppressed by a second tone. The evolutionary importance of such interference is suggested by the presence of adaptations in this auditory system that may function to reduce interference. A major frequency component of the returning echo for this species falls in the band 81–88 kHz. Neurons tuned to this band actually become *more* sensitive in the immediate aftermath of stimulation by the sound frequencies of a typical echolocation call (which are somewhat lower than the echo because of Doppler shift effects).

What sorts of implications do such findings have for vocal communication? When vocalizers are close, and especially when they also produce high-intensity calls, we might expect to find that animals avoid auditory interference by using some alternative to immediate acoustic feedback. Canary-winged parakeets may illustrate this point (Arrowood, 1988). Pair-bonded male–female couples sing antiphonal duets in which the male and female contribute alternating notes; these notes are produced so rapidly and coordinated with the mate's so precisely that they sound like a single, very loud, individual singing (Fig. 4.3). The precision of male–female vocal coordination is indicated by the fact that a pair may

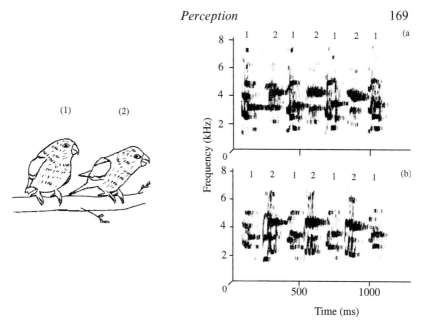

Fig. 4.3. Duetting canary-winged parakeets. The spectrograms show how easily this coordinated duet could be mistaken for the song of a single bird. The male has produced the notes numbered 1, the female has emitted those numbered 2. The two members of the pair are contributing alternating notes to the duet, in a tightly coordinated way. (a) For this recording, the microphone was closer to bird 1, the male, whose notes are consequently more pronounced than those of bird 2. (b) A recording of the same duet, but through a microphone closer to bird 2, the female, whose notes are therefore more pronounced. (Spectrograms and photograph of parakeets courtesy of Patricia Arrowood.)

produce notes at a pace of five to six per second, with neither overlap nor silent gaps between consecutive notes. They often stand within a few centimeters of each other while duetting, even touching at times, with their beaks 1–2 cm apart, and may even be confronting another loudly duetting pair. So, the potential for auditory interference would seem to be high. Consequently, an individual might not be able to time each note by responding to the mate's preceding note. Do they cue each other visually? Perhaps, but not always; sometimes they interrupt mutual allopreening to duet, keeping their heads buried in each other's plumage while singing (Arrowood, personal communication), leaving open only the option of tactile cueing. In addition, they often duet while confronting another pair, where the need to focus on the rival pair might limit visual attention to the mate. So, the coordination may actually be mediated auditorily, but perhaps not on a note-to-note basis, in which auditory interference would be most serious.

There is anecdotal evidence that the members of a pair of canary-winged parakeets may stay coordinated through a process of entrainment (see Section 2.3.2), in which each individual's notes are rhythmically driven by neural oscillators that are activated half a cycle out of phase with each other. Sometimes, for example, an individual produces a few notes in the absence of the partner, because the partner either started singing late or finished early. In those situations, the solitary singer may produce notes rhythmically, leaving silent spaces where the mate would have normally chimed in (Arrowood, personal communication). This persistence of rhythmicity without the partner may be loosely analogous to the persistence of circadian rhythmicity in the absence of the usual environmental entraining cycle. Many species of calling insects and amphibians, such as crickets and frogs, use this entrainment method to adjust their rhythmic calling to that of a vocalizing conspecific (Greenfield, 1994)

However, the parakeets do pick up on some relatively immediate cues; for example, they regularly make coordinated switches between duet types. So, there is some form of on-line monitoring, perhaps of one another's duet types. But, we must not slip into the trap of assuming that all or even most feedback to a vocalizing individual is acoustic. Canary-winged parakeets, for example, may also use visual or tactile cues as a way of coordinating their changes in duet types. Similarly, white-crowned sparrows, described above, not only countersing to the song of a conspecific, but also approach, loom aggressively, and even attack the source of the song (DeWolfe *et al.*, 1989). Attending only to the acoustic dimension of interacting has the potential to 'de-contextualize' communication, deflecting our attention from major sources of insight into communicative behavior (West & King, 1996).

4.2 Motivation

Motivation stands in a two-way causal relationship with perception. The cues picked up by perceptual systems play a role in setting the individual's motivational state. Animals, for example, are often quite prepared to give up feeding to seek refuge when the antipredator calls of conspecifics ring out. At the same time, the motivational state of the individual establishes a context for assessment, focusing the individual's attention on some cues, for example, and deflecting attention from others. So, for example, the extent to which antipredator calls can disrupt an individual's feeding will depend on the implications of the threat of predation.

Female California ground squirrels with vulnerable young respond more strongly to playbacks of antipredator calls, with greater disruptions of feeding and other activities, than adult females without pups (Leger & Owings, 1978).

Motivational systems set the broad regulatory themes of an individual's behavior. As discussed in Chapter 2, these themes are the general life tasks that individuals face – acquiring food, reproducing, avoiding attack by predators and parasites, maintaining social status, and so forth. These broadly defined goals only loosely specify the content of lower levels of behavioral organization, that is, the more specific matters represented in the mode, module, and action levels of behavior systems discussed in Chapter 2 (see Fig. 2.10). When an individual evaluates an event in terms of content, it is judging the relevance of the event not only to major life themes (e.g., avoiding predation) but also to subthemes, modes, and modules (e.g., What exactly should I do about this snake?).

The above levels correspond to what has been called the motive aspect of motivation (Beer, 1982), in the sense of an individual's motives, the specifications in currently active managerial systems about preferred states of affairs. This can be contrasted with the motor aspect of motivation (Beer, 1982), which is most strongly linked to the importance dimension of evaluation, mentioned in the discussion of emotions below. Rather than dealing with content, motor refers to how strongly an individual's behavior is 'turned on,' e.g., how vigorously impelled it is to eat, mate, fight, etc. To speak of such matters in terms of importance, one would ask how much of an individual's time, energy, risk, and attentional budgets the activity can monopolize, e.g., how much injury it will risk in order to engage in the activity.

There is evidence that the motor and motive aspects of motivation affect different dimensions of vocal behavior. For example, both cotton-top tamarins (a small New World primate) and domestic chickens (Fig. 4.4) use specific calls in the presence of food, a particular motive context (Marler *et al.*, 1986; Elowson, Tannenbaum & Snowdon, 1991). Both also vary their rate of calling as a positive function of their motivation to feed (motor aspect), either in the presence of food of varying palatability, or as they vary in level of satiation. Less-sated animals, or animals in the presence of highly preferred food items, call at higher rates. It is of interest that these animals shift the rate of calling but not the type of call that is emitted as their level of motivation varies (Evans, 1997). The type of call seems pretty tightly linked to the motive context of presence or prospects of food, a feature that has led to the conclusion

Fig. 4.4. A male domestic chicken attracting a female to court her by holding a piece of straw in his beak and food calling. (Spectrogram courtesy of Chris Evans; photograph from Peter Marler.)

by some that these calls convey information specifically about food. This topic is discussed under the heading of referential communication in the section on cognition below.

When we speak of an individual's motivational state, we are asking which theme(s) has priority, and how important the individual considers the theme to be. Motivational states are most clearly relevant to management because they deal with what an individual is attempting to accomplish, and how hard it is trying, but they are also relevant to assessment because they focus an individual's attention on the features of the environment most relevant to its current efforts. Cosmides and Tooby (1995) have explained why motivational mechanisms should have this attention-focusing effect. These mechanisms are shaped by selection for their potential to contribute to inclusive fitness. But, the connexions between the proximate details of behaving and inclusive fitness are quite complex and indirect (Thompson, 1986). The fitness consequences of behaving are not immediate enough to guide behavioral decisions, so attentional mechanisms cannot be focused there. And, the consequences that are most relevant depend on the task at hand. The cues for 'moving in the right direction' in a task for acquiring a mate, for example, are often quite distinct from those for deterring a rattlesnake or coyote. So, attention needs to be redirected when motivation changes. From the perspective of communication, this means that the signal that is most salient depends on the individual's motivational state. When an individual is hungry, food-associated signals may be most apparent; when sexually motivated, sex-associated signals may be more salient, and so forth. Remember from Chapter 2, for example, that female laboratory rats are more responsive to the vocalizations of pups when they have been induced into a parenting state by parturition and care of young, than when they have not recently had that experience (see Section 2.3.1).

Is there additional evidence that an individual's response to vocalizations depends on its motivational state? Yes. Parenting often involves responding to cues more indirectly relevant to offspring, such as the acoustic signals of predators (Swaisgood, 1994). As described in the Prologue and Chapter 2, rattlesnakes are an important source of predation on California ground squirrel pups, but not on adults. Adults confront these snakes, and maternal females spend more time in this activity than nonmaternal females and males, neither of whom contribute much to the care of pups. Rattlesnake confrontation by adults can be aggressive enough to induce the snake to rattle at the squirrel, and this sound incidentally includes cues about the degree of risk that the confronting squirrel faces. In playbacks of rattling sounds, maternal females respond more strongly, and differentiate more finely among these acoustic risk cues, than nonmaternal females and males (Fig. 4.5).

4.2.1 'Intentional' signaling

Exploring the motive aspects of motivation provides insights relevant to a recently visible issue in the study of animal communication: the interest generated by game theory in the extent to which animals signal their 'intentions' (e.g., Hauser & Nelson, 1991). This issue was touched upon in Chapter 2, noting that from a game-theoretic approach, intention is more a statistical than a cognitive concept. That is, signals about intentions are defined simply as signals that are statistically predictive of the signaler's subsequent behavior; no higher-order cognitive abilities are

mation about intentions and thus should be selected to ignore all ritualized displays which seem to proffer information regarding motivation. Managing individuals, on the other hand, are assumed to be less behaviorally constrained by their motivational states than by their RHP; behaving in a way that is not consistent with one's motivational state is thought to be quite feasible. In other words, it is generally assumed that the signal 'I will attack' is more easily faked than 'I am risky to pick a fight with' (Maynard Smith, 1982, p. 3). That such assumptions are not necessarily valid can be illustrated by exploring two very visible examples from the animal conflict literature – the meral spread by mantis shrimp (Dingle, 1969), and the state of musth in male African elephants (Poole, 1987).

Mantis shrimp provide us with one of the more compeling examples of signals 'about' aggressive intent, albeit not involving vocal communication. They defend their home cavities against repeated intrusions by neighbors, in part through the use of a signal called a meral spread in which the dangerous claws are held up in display at the intruder (Steger & Caldwell, 1983; Caldwell, 1986; Adams & Caldwell, 1990). The effectiveness of this display in repeling intruders is maintained by following it with a potentially lethal smash with the claws. Such apparent Pavlovian conditioning of neighbors is intensified at a critical time, just prior to each shedding of the exoskeleton, during which the individual is very vulnerable because it has neither armour nor weapons. The result is that the meral spread remains effective as a bluff through the molt (15–20 percent of the individual's life), when a smash cannot be delivered. The feasibility of bluffing is further enhanced by the fact that the state of molt is not detectable either visually or chemically.

So, RHP can be as covert, and as abruptly changeable as motivation is often assumed to be. Assessing individuals are constrained in their ability to detect the precipitous drop in RHP that comes with each molt, not only because the molt is so cryptic, but also because of the danger involved; a challenged mantis shrimp can deliver a lethal smash during the 80–85 percent of its life that it has an exoskeleton. Other examples of rapid reversals in RHP can include physical exhaustion from fighting (Clutton-Brock & Albon, 1979), the breakage of horns and antlers used in combat (Geist, 1966), and changes in the body temperatures of ectotherms (Rowe & Owings, 1990). Similarly, additional cryptic determinants of fighting ability include condition (e.g., Prins, 1989), experience (e.g., Berger, 1986), and skill (e.g., Berger, 1981). Such factors set limits on the extent to which assessment systems can force management systems to be sources of reliable cues.

The state of musth in African elephants has also been presented as an 'announcement of intent' (Poole, 1989a). Musth is a state of heightened aggressiveness and sexual activity in male elephants, analogous in many ways to the state of rut in other male mammals. Musth is accompanied by a dramatic increase in testosterone levels (Poole *et al.*, 1984), as well as other characteristic physical and behavioral manifestations (Poole & Moss, 1981; Poole, 1987), including visual displays, odor signals, and a low-pitched vocalization called a musth rumble (Fig. 4.6). Unlike rutting in deer, musth in male elephants is asynchronous, i.e., different males may be in musth at different times of the year. Although body size is a significant determinant of rank, and thus of mating success, small males in musth are dominant over larger males in nonmusth. But why? Poole (1989a) argues that an estrus female represents a more valuable resource to a sexually-more-active male in musth than to a sexually-less-active male out of musth; thus Poole interprets musth as a signal of a male's aggressive motivation to fight for a valuable resource.

But if musth contributes so directly to a male's reproductive success, and if intentions are easy to fake, why not always signal musth? If a male could stay in musth longer, might he not increase his fitness? A partial explanation may relate to changes in male body condition with musth. Because they eat less, walk more, and have elevated metabolic rates relative to nonmusth individuals, males in musth rapidly lose condition. In Poole's study, weight loss was significantly correlated with duration of musth, and two males that had the longest duration of musth one year failed even to enter musth the next.

The evidence discussed above suggests that musth is constrained by the physiological ability to sustain it. A male elephant's aggressive motivation, then, appears tied to his physical condition. Indeed, musth may be viewed as a suite of physiological adjustments that simultaneously increases fighting ability and lowers aggressive thresholds. In support of this argument are the known parallel effects of testosterone: not only does testosterone facilitate aggressive motivation, but it also increases muscle mass (at least in the short term), promotes growth of motor neurons, mobilizes energy, and enhances cardiac function (Dixson, 1980; Mainwaring, Haining & Harper, 1988). (For those who remain doubtful about the effects of hormones on physical ability, consider the prohibition of the use of androgens by athletes because of the unfair advantage in physical prowess that these steroids provide.) In addition,

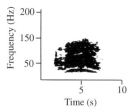

Fig. 4.6. The state of *musth* in mature male African elephants is similar to the state of *rut* in other male ungulates. While in *musth*, elephants produce olfactory signals, such as the exudate from the temporal gland and the dribbling of urine from the penis, wave their ears in a conspicuous visual display, and emit a very low-frequency vocalization called a *musth rumble* (see spectrogram). (From drawings courtesy of Joyce Poole.)

if we assume that aggressive motivation is tied to expected probability of winning, musth may confer on an individual a preliminary self-assessment of superior fighting ability, thus reinforcing its aggressive motivation (Hamilton & McNutt, 1997). Musth may therefore be interpreted not as a simple announcement of intent, but as a state of altered RHP and motivational priorities which includes cues about both of these variables. Indeed, displays used during the period of musth may be designed merely to emphasize the elephant's heightened fighting ability. To the extent that fighting ability and intentions are correlated, the state of musth may also afford cues regarding broad motivational state (i.e., probability of escalating) and subsequent changes in the use of relatively discrete displays which predict attack.

This analysis illustrates the utility of the closer scrutiny of proximate processes prescribed by an A/M approach: fighting ability and aggressive motivation may not be as independent as theory has assumed. Although similar points have been made before (Enquist, 1985; Markl, 1985; Turner & Huntingford, 1986; Adams & Caldwell, 1990), the caveat seems not to have had the impact it merits (Hamilton & McNutt, 1997). If an individual's motivation and ability to fight are correlated, it may be difficult or impossible to disentangle signals that capitalize on assessment of intent from those that exploit assessment of RHP. Although there may be utility in making a conceptual distinction between intentions and fighting ability, it does not necessarily follow that animals should have evolved separate signals for these two components of dangerousness (e.g., Neil, 1983). Constraints such as those proposed for African elephants may link the two variables causally, thus limiting the ability to capitalize on assessment of one without influencing assessment of the other. Indeed, from the perspective of assessment, all that matters is the product of RHP and motivation, i.e., how much RHP the adversary is motivated to use (Parker & Rubenstein, 1981; Markl, 1985). The managerial goal is only to deter the opponent by convincing him, by whatever means, that the signaler may inflict damage.

Finally, yet another dichotomy between RHP and intentions established by game theorists seems to be blurred by a consideration of proximate mechanisms. It is generally assumed that increases in RHP entail a cost, whereas intentions can be altered without any intrinsic cost (e.g., Maynard Smith, 1982). Thus signals indicating heightened aggressive intent are considered to be relatively inexpensive and therefore easily bluffed. However, to the extent that (1) motivational changes are mediated hormonally, and (2) hormones produce real physiological costs, this game-theoretical assumption that motivation entails no costs appears unrealistic.

4.2.2 *Pumping up: the self as target*

As discussed earlier, the behavior of 'lying about one's intentions' is generally considered to be easy because it entails little or no cost in terms of constraining underlying motivational states. It is assumed that animals can 'feel one way and act another.' Zahavi (1982) and Bond (1989b) have argued that behavior cannot be so easily decoupled from motivation because specific behavioral activities are often contingent upon particular physiological/motivational states. Bond argues that gradual escalation in conflict occurs as a result of self-regulatory processes in which engaging in a behavior provides feedback to motivational/physiological systems, thus priming them for the incipient need for heightened physical activity. If such acts are necessary precursors, providing adjustment of physiological/motivational systems supporting aggressive behavior, then they need not be viewed as 'signals' at all in the sense of having been specialized to influence the behavior of others. Indeed, the role of these activities as sources of information may be better understood from the perspective of assessment than of management. Since these acts are necessary precursors to aggression, these unritualized behaviors leak information upon which active perceivers can capitalize.

This scenario provides an excellent example of a fundamental point in an A/M approach: the proximate bottom line in management is self-regulation. Although animals often meet their own needs by managing the behavior of other individuals, they just as often meet their own needs by operating directly on themselves. (For example, the young white pelican depicted in Fig. 2.3b regulated its body temperature initially only by squawking to evoke parental incubation activities, but not long after hatching was able to shiver as a more direct means of thermoregulation.)

Two examples support the idea that individuals may prepare themselves for important activities in part through self-stimulation. First, Hollis (1984; 1990) used a Pavlovian paradigm to condition territorial male blue gouramis to expect an intruder following the presentation of a light. This enabled these fish to prepare for the incipient intrusion by engaging in early aggressive signaling such as frontal display and threat posture (Hollis suggests the possibility that such preparation may be accompanied by an anticipatory release of testosterone). Subsequently, they confronted rivals in a 'pre-escalated' mode, winning more fights and delivering significantly more bites and tailbeats. So, male gouramis can gain a distinct advantage over interlopers if they can learn to respond to events which predict impending encroachment of rival males. Such predictive cues would be a part of natural interactions as well.

Cheng's work with ring doves deals with the *coo* vocalizations used by both sexes during courtship (Cheng, 1992). Reproduction in this species involves a cascading series of changes in male and female. Males typically initiate courtship by bowing and *cooing*, which is followed by the male's *cooing* over prospective nest sites. The female gradually comes to join the male in *cooing* over the prospective nest site, and ultimately engages in a long stint of solo nest-*cooing*, before the two join forces in the construction of a nest. When nest building reaches a threshold level, hormonal changes are triggered in the female which culminate in ovulation and copulation. What role do the female's *coos* play in this process? It has

seemed reasonable to identify the male as the target of these vocalizations. However, muting the female in several different ways leaves the
male's courtship activities relatively unchanged, but blocks the hormonal
changes in the female leading to ovulation. And, playbacks of *coos* to the
female restore those changes, especially when the vocalizations used are
her own (Fig. 4.7). Further playback studies indicate that the female *is*
the target of the male's calls, but they have their effects on the female by
stimulating her to *coo*, which in turn induces her to ovulate through a
process of vocal self-stimulation.

The point of the preceding paragraphs is that evolutionary theorizing
without dealing with the properties of the proximate mechanisms
involved is likely to lead to mistaken or oversimplified models.
Adaptive behavior is accomplished through mechanisms whose properties are major determinants of what works best.

4.2.3 Refining the concept of intentions

Poole's interpretation of musth also reveals the potential confusion that
can arise as a result of defining 'intention' as any behavior that predicts,
at least probabilistically, what an animal will do in the future. This definition may incorporate a variety of motivational processes. Thus it may
be useful to distinguish among processes which operate over different
time frames or at different organizational levels (see Fig. 2.10; see
Dennett (1983) for a different, more cognitive, formulation of levels of

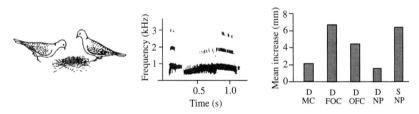

Fig. 4.7. Two ring doves courting at a nest site. The *coo* vocalization, depicted in
the spectrogram, is emitted repeatedly by both male and female during courtship,
and the female *coos* repeatedly in solo during the latter stages of nest-site selection. Playback studies demonstrate that the female's own *coos* are more effective
in stimulating growth of ovarian follicles than any other source of vocal input. D
= female devocalized; S = female sham devocalized; MC = male *coos* played
back to female; FOC = female's own *coos* played back to her; OFC = other
female's *coos* played back to female; NP = no playback to female. (Spectrogram
courtesy of M.-F. Cheng; bar graph from Table 1 of Cheng, 1986.)

intention). Motivation is often thought of in terms of states (such as hunger and the phenomenon of musth; the molar left end of Fig. 2.10) which influence the response probabilities of the organism (e.g., Wise, 1987). Such states are generally characterized by a certain amount of inertia. The concept of motivation, however, also subsumes moment-to-moment decision-making processes regarding what an animal will do next (the more molecular right end of Fig. 2.10, i.e., the everyday way of thinking of 'intentions').

In our view, the absence of a clear distinction among these organizational levels of motivation has given rise to some confusion regarding predictions from evolutionary analyses. Putative 'signals of intent' include both signals that say 'On my next move I will attack' and those which say 'I will persist in an escalated contest.' The former applies to the molecular level of motivation, but the latter may relate to molar adjustments in motivational state which prepare for and thus predict future actions. Compare, for example, demonstrations that an animal's next move can be predicted by its previous behavioral act (Dingle, 1969; Bossema & Burgler, 1980; Nelson, 1984; Waas, 1991), with evidence of the possibility of longer-term predictions (Riechert, 1978; Turner & Huntingford, 1986; Barlow, Rogers & Fraley, 1986; Poole, 1989b). Discussions of signals of intent rarely differentiate among these different kinds of intentions. Maynard Smith (1982), for example, appears to refer interchangeably to signals which indicate what an animal will do next, and to signals that indicate willingness to persist. As a result, empirical tests often do not clearly distinguish which prediction is being tested, in part because they do not address temporal/organizational dimensions of motivation.

The A/M approach deals explicitly with different levels of proximate influence on behavior. Moment-to-moment intentions, therefore, would be embedded in broad motivational states and in two-way interaction with them. Thus a shift in motivational state implies a changing distribution of motives, thereby potentially influencing behavior which predicts both immediately forthcoming actions and the patterning of behavior over a longer time scale. Such proximate considerations may prove useful in refining ultimate formulations because different levels of organization may be subject to different kinds of constraints. For example, it might be found that it is easier to fake or conceal intentions (what an organism may do next) than motivational state (a systemic adjustment reflecting preparation for certain kinds of activities, e.g., preparedness to escalate).

4.3 Emotion

Emotions ensure that high-priority life themes are taken care of. Positive emotions are sources of feedback that sustain the activities that produce them. Negative emotions are sources of feedback that disrupt ongoing activities, redirecting efforts to high-priority themes. In both cases, emotions provide partial answers to that most Darwinian of questions, 'What does this mean for me?' (Zajonc, 1980). Evaluation of the significance of events and conditions to the assessing individual is a critical aspect of the assessment process. Such evaluations can be said to have two general dimensions – importance and polarity (positive to negative). Emotional involvement increases as importance increases, but the specific nature of the emotion depends in part on whether the emotion is positive or negative.

An individual's behavior depends in part on these evaluative reactions to its experiences. Many of these evaluative reactions are linked to the well-being of the animal; painful inputs, for example, often involve tissue destruction, which can be detrimental to the animal's well-being. Animals exhibit many signs of finding painful stimulation aversive. They often work hard to limit painful stimulation, not only escaping from sources of such input, but also anticipating pain and avoiding activities and places previously associated with it (Bolles, 1970). However, not all evaluative processes have to do with well-being *per se*. For example, male rats find ejaculation highly rewarding (Agmo & Berenfeld, 1990), but there is no reason to believe that ejaculation contributes to a male's well-being. Indeed, the prerequisites and consequences of ejaculation, at least in the usual way through copulation, can be painful and detrimental to well-being. Males may exhaust and injure one another during sexual competition, females may injure males during copulation, and the surge of testosterone that can follow ejaculation can suppress the immune system (Herndon, Turner & Collins, 1981; Zuk, 1994; Boellstorff *et al.*, 1994).

The key to understanding this apparent paradox is the logic of natural selection, which favors emotional systems that contribute to the fitness of individuals, selecting systems that promote well-being only to the extent that well-being contributes to fitness. Nevertheless, these evaluative systems exist, and need to be dealt with. For example, when animals need to contend with the possibility of pain in order to accomplish something adaptive, there are analgesic systems that are biochemically activated, for example through the release of internally produced opium-like substances

(Fanselow, 1991). The aversive motivational consequences of painful stimulation are inhibited in this way. Otherwise, the powerful motivating effects of pain might interfere with adaptive behavior.

As noted, these positive and negative evaluative reactions can become attached to cues of impending reward or punishment. Male rats, for example, are attracted to stimuli associated with locations where they have ejaculated (Agmo & Berenfeld, 1990) but avoid stimuli associated with locations where they have experienced painful stimulation (Bolles, 1970). So, human experimenters and animal trainers can manage the behavior of rats in part by using such signs, suppressing behavior by following it with stimuli associated with pain, or reinforcing behavior by following it with stimuli associated with positive consequences. We suggest that animal signals often work in an analogous way, i.e., through their association with positive or negative consequences. Of course, learning may or may not be involved; the 'discovery' of the association between the signal and positive or negative events no doubt usually involves natural or sexual selection. But, the effects are similar; vocal signals should affect the behavior of others in part by evoking emotional states and thereby motivating strong behavioral reactions. Indeed, the presence of these evaluative systems should create selection on managerial activities for effectiveness in capitalizing on them.

What is the evidence that vocal signals achieve their effects on targets by influencing their emotions? There is a substantial history of interest in the extent to which emotions play a role in the *emission* of signals (e.g., Marler, 1984). Surprisingly, however, substantially less attention has been paid to the question of the emotional impact of signals on targets (Klinnert *et al.*, 1983; Scherer, 1992; Owings, 1994). Nevertheless, the beginnings of a story can be pieced together. As discussed in the Prologue and in Chapter 2, the prosodic features (melodies) of human speech work through the emotional systems of infants. A few months after exhibiting such emotional responses, these infants develop the ability to use social referencing during times of uncertainty, such as when they have been confronted with an unfamiliar object or person (Klinnert *et al.*, 1983). Social referencing involves taking cues from the emotional expression on the mother's face. Working hard to keep their mother's face in view, infants engaging in social referencing mirror the mother's emotions in their own facial expressions, and are more likely to make contact with stimuli about which they are uncertain if the mother's facial expression indicates a positive rather than a negative emotion. Such social processes are not unique to humans. Both rhesus monkeys and

European blackbirds catch the concern of conspecifics engaged in responding vocally and in other ways to a predator. If the object of the conspecific's concern is visible, the observer will develop a long-term fear of it (Vieth *et al.*, 1980; Mineka & Cook, 1988). The example of laughter by humans, discussed below, provides an especially compeling case.

The use of animal models of stress and anxiety indicates parallels between the sociophysiological mechanisms of human and nonhuman emotional responses. For example, when laboratory rats are subjected to electrical shock to the feet, they respond by vocalizing, jumping, struggling, defecating, and urinating. When other rats are allowed to observe these reactions, without experiencing pain themselves, the observing rats show signs of catching some of the stress that the shocked rats are experiencing (Kaneyuki *et al.*, 1991). Within ten minutes, the observing rats' adrenal glands release corticosterone, a sensitive hormonal index of emotional response to stress. After 30 minutes of such observations, chemical changes take place in the medial prefrontal cortex, a portion of the brain believed to play a role in control of negative emotional states, such as anxiety or fear. These changes in the prefrontal cortex are blocked by diazepam (Valium), a drug prescribed to humans for the reduction of anxiety.

It is reasonable to ask whether the construct of emotion adds anything to our understanding beyond what the concept of motivation provides (e.g., Fridlund, 1994). Berridge's (1996) work on the mechanisms underlying food reward indicate that both constructs are essential. Different behavioral assays are required to measure the two, and different systems of brain structures and neurotransmitters underlie them. It has been traditional to treat an animal's desire for food and its affective evaluation of food as essentially the same process. That is, animals are assumed to want what they like, and like what they want. However, Berridge's behavioral research indicates that wanting (motivation) and liking (affect/emotion) are separable systems, that do not always correspond in their impact on behavior. Wanting involves an active, instrumental 'reaching out' to reward, paralleling in many ways the ethological concept of appetitive behavior (and the levels of organization more toward the left side of Fig. 2.10). Wanting is assessed via a variety of instrumental procedures, many of which measure how hard the individual will work to gain access to the reward. Liking, on the other hand, parallels in many ways the ethological concept of consummatory behavior (and the levels of organization more toward the right side of Fig. 2.10). Liking is assessed by recording immediate behavioral reactions to the rewarding or aversive input. Laboratory rats, for example, display distinctively different patterns of facial and forelimb actions to sweet versus bitter tastes that are similar in many ways to the reactions of human infants to the same flavors. Sweet tastes elicit rhythmic mouth movement, distinctive tongue protrusions, and prolonged lapping at the source. Bitter flavors evoke mouth gaping, rearing, shaking the head, facial wiping with the forelimbs, and shaking the paws. Most experimental manipulations that affect human perceptions of palatability also influence these affective reaction patterns by rats.

4.3.1 Emotional contagion

Peter Sellers was a master of cinematic comedy. As the son and daughter of the Owings family were growing up during the 1970–80s, the family enjoyed sitting down together to watch videos of Inspector Clouseau entangling himself in and miraculously extricating himself from yet another series of hilarious mishaps. Of course, it was not always possible to assemble the entire family, and one participant proved to be critical for maximum enjoyment; Ragon, the big brother, was the real connoisseur of the Pink Panther films. He would roll on the floor, almost helpless with laughter, and his laughter was contagious. The whole family laughed more and enjoyed the films more with the added stimulation of Ragon's laughter.

This infectious effect of Ragon's laughter is not unique. Laughter, a distinctive human vocal and visual signal (Fig. 4.8), typically induces others to laugh; and it is more than just the vocal pattern that is contagious. Laughter is also a means to spread a positive emotional state to

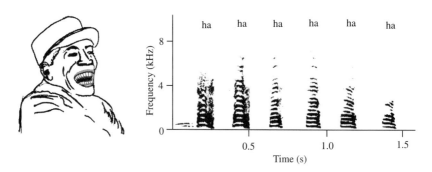

Fig. 4.8. Laughter, a species-typical human vocalization and facial expression. (Spectrogram courtesy of Robert Provine.)

others (Provine, 1996b). Broadcast radio discovered and applied this biological fact early on (Provine, 1996a). Ed Wynn's first live comedy performances on radio in the 1920s were seriously disrupted without feedback from a laughing audience, so he initially recruited the stage crew, and subsequently began to use 'laugh tracks' (recordings of laughter). Ultimately, broadcasters discovered that audiences enjoyed programs more with the laugh tracks, and continued to incorporate them for that reason. From the 1950s to the present, many television comedies have also exploited this playback procedure as a way of enhancing program ratings.

What are the processes underlying emotional contagion? Three general categories have been identified in humans – conscious cognitive processes, unconditioned emotional responses, and conditioned emotional responses (Hatfield *et al.*, 1994). (1) Conscious cognitive induction would include vividly imagining oneself in an emotionally charged situation currently being described by a companion. (2) Unconditioned emotional responses can be illustrated by the emotional responses of human infants to our distinctive ways of speaking to them, called motherese (see Prologue and Fig. 2.4). There is evidence that such processes are based in part on mimicry of motor patterns. Human infants, for example, mimic the facial expressions of adults from an early age, and the adoption of those expressions has the potential to induce the associated emotional state (Ekman, 1992; Hatfield *et al.*, 1994) (3) Conditioned emotional responses are readily induced through Pavlovian conditioning procedures in both humans and nonhumans (Fanselow, 1991). A common method is to pair electrical shock to a rat's feet with an initially neutral stimulus such as a tone; the rat quickly comes to anticipate the shock when it hears the tone, taking behavioral and physiological steps associated with the state of fear that can be instrumental in avoiding the shock. Actively assessing individuals should discover and use such predictive cues all the time in their natural circumstances, responding to the call of a conspecific in part on the basis of the behavior associated with that call type during past interactions with that individual (Hollis, 1990; Owren & Rendall, 1997).

4.3.2 An affect-conditioning model

Conditioned emotional responses have been proposed to play a central role in nonhuman primate vocal communication (Owren & Rendall, 1997). This model deals relatively little with the details of emotion, concentrating instead on the importance of individuality in vocal structure for natural Pavlovian conditioning processes. Since this form of learning is a basic cognitive process (Rescorla, 1988), it provides a good topic for bridging to the next section on cognition.

The affect-conditioning model from which this proposal was derived is consistent with an A/M approach. The model calls our attention to proximate issues other than the information conveyed by signals, including the fundamental role of emotional processes in communication, and the decisive importance of attention to call structure. The result is a novel way of thinking both about the significance of individual variation in vocal structure, and about the acoustic structural sources of individuality.

Some vocalizations of nonhuman primates vary among individuals (e.g., the *coo* calls of rhesus monkeys), and these cues about individuality make a difference in playback studies. In contrast, other calls by the same species (e.g., noisy screams) show little evidence of individuality (Rendall *et al.*, 1996; Rendall, 1996). From an informational perspective, the natural response to such data has been to focus on the differing information these calls make available about the identity of the caller. In contrast, Owren and Rendall begin their inquiry with the more basic, pragmatic position that signals are used to influence the behavior of targets; they then explore the structural bases of vocal individuality, and the significance of variation in individuality for emotional conditioning in communication.

Owren and Rendall note that mechanisms of Pavlovian conditioning are very widespread in the animal world, and that these mechanisms almost certainly play a role in social interactions. One of the most compelling demonstrations of the social role of Pavlovian conditioning can be found in the work, described above, on the use of the meral spread display by mantis shrimp (but Caldwell and his colleagues do not use the term 'conditioning'). The individually distinctive cues associated with a specific mantis shrimp's meral spread are paired with aggressive lunges and smashes, allowing that individual to rely on the meral spread alone as a conditioned stimulus during subsequent encounters with its neighbors. Although this might result in enhancement of the impact of meral spreads by all individuals, the fact of dominance relationships indicates that the conditioning effects also accrue selectively to the individual signaler/ smasher.

Owren and Rendall note that the social interactions of nonhuman primates also involve the pairing of (vocal) signals with negative and positive emotional consequences. For example, in many primate species, former opponents reconcile with one another after fights, engaging in friendly interactions at higher rates following aggressive interactions than at other times (Silk, Cheney & Seyfarth, 1996). Among adult female baboons in the Okavango Delta of Botswana, *grunt* vocalizations are regularly used by the initiator of reconciliation, and these calls seem to reduce the target's subsequent apprehension about aggression from the initiator (Cheney, Seyfarth & Silk, 1995). Owren and Rendall, in this case, suggest that the association between grunts and friendly interactions has given grunts the power to serve as a 'safety signal,' evoking a conditioned positive emotional reaction in the target individual. In fact, the grooming interactions that can follow grunting have the effect of inducing the release of endogenous opium-like neurochemicals, which are components of a mechanism that mediates the rewarding effects of friendly social interactions (Keverne, Martensz & Tuite, 1989; Martel *et al.*, 1995) Among rhesus monkeys, fighting can also be a very vocal affair. Individuals who are confident of their ability to win a fight pair their threats and aggression toward subordinate individuals with a roar vocalization (Rowell, 1962). Such pairing could make the dominant monkey's roars an aversive conditioned stimulus for the individuals who have experienced the juxtaposition of roar and aggression. Owren and Rendall conclude that the well-understood processes of Pavlovian conditioning, which are known to be effective in managing the emotional states of target individuals (Fanselow, 1989), provide a mechanism for the impact of individually distinctive primate vocalizations that is much more tangible than the metaphor of information exchange. These conditioned effects would be specific to the history of interaction between the individuals of interest, and would function in addition to the unconditioned effects expected from the evolutionary processes upon which, for example, motivation–structural rules are based (see Chapter 3).

The affect-conditioning model is useful in accounting for variation among primate vocalizations in the availability and form of individual cues. Two sources of such cues can be identified – one having to do with the transfer function of the vocal tract (Fig. 4.9), the other with distinctive temporal patterns. Transfer functions have to do with the shape and size of the supralaryngeal vocal tract, that is, the portion of the throat, mouth, and nasal cavity into which sound is projected by the vocal folds of the larynx. Even where the vocal folds of different individuals produce the same source sound, the different shapes and sizes of the tracts of different individuals can generate different resonance patterns, reinforcing and attenuating different frequency components (creating formants),

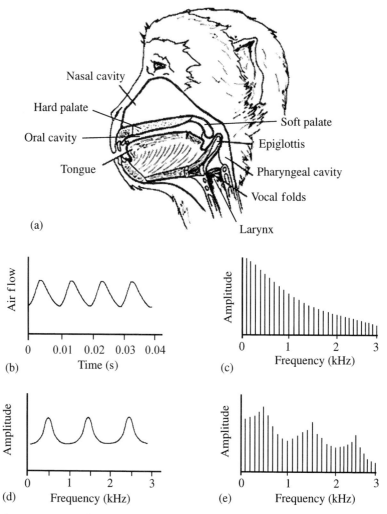

Fig. 4.9. Sources of variation in the structure of the vocalizations of primates, and of mammals in general. (a) A midsaggital drawing of the head of a rhesus monkey. The sources of vocal sounds are the vocal folds ('cords') in the mouth of the larynx. These sounds are projected into the supralaryngeal vocal tract, whose resonant properties filter the sound, by amplifying some frequency components of the source sound and attenuating others. This *source-filter* model of vocal structure is illustrated in b–d. (b) The vocal folds are opening and closing 100 times per second, yielding (c) a sound with a fundamental frequency of 100 Hz, and harmonics at integer multiples of 100 Hz, which decline exponentially in amplitude with increasing frequency. (d) The supralaryngeal vocal tract amplifies certain source-sound frequencies, creating *formants* (regions of emphasis) around 0.5, 1.5 and 2.5 kHz. (e) These filtering effects (the *transfer function*) yield a final sound structure with a quality strongly influenced by both source and filter contributions. (Courtesy of Michael Owren.)

thereby making the calls of different individuals distinctive. This source of individuality of calls is not very flexible for a given source sound (in contrast to humans); it places an inevitable signature on calls. But, there is variation in the conspicuousness of such transfer-function signatures. They become more apparent where more sound frequencies are available over which to distribute the signature. These conditions are met by harmonically structured tonal sounds of lower fundamental frequency (*sonorants*, e.g., the *coo* in Fig. 4.10), and by impulsive sounds consisting of broad-band noise (*gruffs*). Transfer-function signatures become less apparent, both as the fundamental frequency of harmonically structured tonal sounds increases (such *shrieks* provide fewer harmonics over which to 'write' the signature), and as sound intensity increases beyond a certain point (such *screams* are produced with the mouth wide open, which reduces the filtering capacity of the vocal tract, e.g., the noisy *scream* in Fig. 4.10). The second source of individuality – distinctive temporal patterns of calling – can be used even for those sounds not individually stamped by the vocal tract. Temporal   patterns can take a variety of

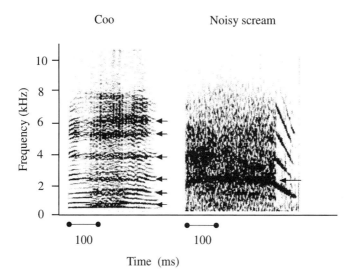

Fig. 4.10. *Coos* and noisy *screams* of rhesus monkeys. The *coo* illustrates the many formants (identified by the arrows) available as cues to the identity of the caller. These transfer-function cues are especially abundant here because the low fundamental frequency and moderate amplitude of the *coo* provide many harmonics to reveal the many points of resonance of the vocal cavity. The noisy *scream* reveals many fewer formants, because of its higher amplitude and fundamental frequency (the latter revealed by the arrow). (Courtesy of Michael Owren.)

forms, including frequency or amplitude contours, distinctive sequences of different call types, and so forth. These should be less obligate cues than those arising from transfer functions, and might therefore vary more in their availability.

Predictions of the model are largely untested, as yet, but provide a rich basis for future empirical test. Predictions by Owren and Rendall include the following. (1) If the pattern of formants in primate calls is as important as the affect-conditioning model suggests, we might expect to find that corresponding auditory systems are particularly sensitive to the positioning of formants. (In an A/M approach, this should be true because signals are said to be selected to capitalize on salient assessment dimensions.) Japanese macaque monkeys are indeed very sensitive to shifts in the positioning of formants, much more so than to changes in the fundamental frequency of sounds; this is just the reverse of the pattern exhibited by humans, who detect shifts in fundamental frequency much more acutely. (2) Where cues about individuality are unavailable in the structure of single calls, we might expect two patterns. First, we should find evidence of compensation for this loss via repetition of calls so that individuality can be replaced via the cross-call patterning of vocalizing. Consistent with this prediction, shrieks and screams, which lack rich formant structure, are typically emitted in bouts, rather than as individual calls (Gouzoules *et al.*, 1984; Gouzoules & Gouzoules, 1989). Second, the loss of conspicuous transfer-function-based individuality should be offset by some compensatory gain. In the case of shrieks and screams, the compensation appears to come in the form of strong unconditioned effects on targets. Shrieks and screams are certainly noxious sounds, in part because they are the loudest vocalizations, at least in the vocal repertoires of macaque monkeys. But they also share other structural features with crying by human babies – a set of vocalizations widely acknowledged to be aversive (Gustafson & Green, 1989) – including sound energy at relatively high frequencies and dysphonation (irregular movement of the vocal cords, resulting in a poorly defined fundamental frequency, as in the noisy *scream* of Fig. 4.10, relative to the *coo*). The repetition of these sounds also contributes to their noxiousness, in addition to restoring cues about individuality. So, shrieks and screams should be effective in deterring continued aggression by the adversary of the subordinate individual.

4.4 Cognition

As discussed in Chapter 2, that we use 'cognition' broadly to refer to all input-processing activities beyond perception, including storing, retrieving, computing, and integrating as a basis for behavioral decisions (consistent with Mason, 1986; Dyer, 1994). According to this inclusive definition, even simple processes like Pavlovian and operant conditioning are cognitive processes. Consistent with the concern in this chapter about integrating management and assessment, it is also argued below that most or perhaps all cognitive mechanisms used in animal communication are parts of action systems, that judge objects and events in terms of what to do about them, not in terms of some abstract representation isolated from its functional context.

Cognitive scientists distinguish between 'data-driven' and 'conceptually driven' cognitive processes (Cohen, 1990). Data-driven processes are initiated and guided by cues coming in from the external world and currently being received by the sense organs. Such processes are perhaps most evident when external events are high-priority ones, such as when a source of danger threatens an individual or its offspring, or a very attractive, sexually receptive potential mate has just presented itself. At those times, current stimulation appears to be a dominant determinant of an individual's behavior. Conceptually driven processes are, for our purposes, more appropriately labeled 'state driven,' because of our concern with the role of motivational and emotional states in addition to cognitive ones in assessment.

The role of internal states is revealed in part through selective attention, and can be illustrated with the phenomenon of latent learning from the animal-learning literature (Simon, 1994). When a hungry rat negotiates a maze that contains food and water in different locations, it learns in relatively few trials how to travel directly to the food location with few wrong turns. However, if it is subsequently made thirsty instead of hungry, it will reveal that it also learned the location of the water, because it improves even more quickly in its travel to the water than it did originally for the food reward. Such findings were treated as evidence that learning, e.g., the location of water, can occur without motivation to acquire water. The learning simply remains latent until thirst motivates expression of it. However, subsequent research proved that this interpretation was too simple. A more modern view demands appeal to processes of attention. Rats learn whatever they *attend* to, but the direction of their attention depends on their motivational state. If hungry, but not too hungry, rats note and remember where water was detected during their search for food. But if highly food deprived, their attention is not attracted by anything but food; water goes unnoticed and its location remains unlearned until it is exposed to the maze in a state of thirst.

The example of latent learning illustrates how cognitive processes such as learning are typically driven not separately by data or internal factors, but by both. An animal's cognitive representation of a maze is determined in part by its experience with the maze, but the details that are highlighted in the map depend in part on internal factors such as its motivational state. From an A/M perspective, an individual's cognitive processes could be said to depend on its current regulatory problems. A hungry rat has experienced a disturbance in the regulation of its nutritional state; a thirsty rat's water balance is disrupted.

4.4.1 Context

Animals are regularly faced with the question 'What should I do next?' and the answer, according to what we have said so far, is 'It depends on what the current regulatory problems are.' Superficially, this sounds straightforward; individuals simply need to identify their current regulatory problems, and then the next step should be self-evident. But life is more complicated than that. It is not always clear what problems an individual is facing. A predator may be nearby but undetected, because it is hiding in ambush. Or, another member of the group may be surreptitiously attempting to usurp the individual's position of social status. Even when such problems are uncovered, the solution may not be clear. If an individual discovers a social climber, it will usually be confronted. But, how assertive should the individual be? It depends on the climber's aggressive motivation and ability (Hamilton & McNutt, 1997). One source of cues on those matters is the structure of the vocalizations that the climber emits (e.g., Davies & Halliday, 1978). But, vocalizations may not always be the most reliable source of cues (Markl, 1985), because they have been shaped by selection for effectiveness in managing the behavior of opponents, including bluffing when necessary.

Conditions such as those described above are important sources of selection for very active assessment. Problems are not always immediately evident, so active efforts may be required to uncover them. Once problems are identified, the need for a reliable foundation for behavioral decisions favors the supplementation or even replacement of cues from individual signals with contextual cues. The complexities of social assessment and interaction place a premium on the cognitive ability to integrate cues from multiple sources.

What are the contextual sources available for more precise assessment? One important source consists of the patterning of signaling, a level of organization that is independent of the structure of individual signals. The red deer stags of Scotland make use of the rate of roaring as an assessment cue during conflict with other males for sexual access to groups of females (Clutton-Brock & Albon, 1979; Clutton-Brock, Guinness & Albon, 1982). A contest begins when a challenger approaches to within 200–300 metres of a harem-holding stag. The two usually roar at each other for several minutes, and the challenger typically then withdraws. If, however, the challenger approaches even more closely, the adversaries exchange an additional set of roars and then may progress to a parallel walk, in which the stags walk tensely, side by side, perpendicular to the direction from which the challenger approached. About half of the parallel walks escalate to actual fights, in which individuals lock antlers, pushing, twisting, and even stabbing the opponent if an opportunity arises. The consequences of losing a fight can be serious; about 23 percent of stags over five years of age show some sign of injury during the rut each year, and up to 6 percent are permanently injured. So, accurate assessment of opponents is important. Playback studies indicate that males attempt to roar more rapidly than their adversary in roaring contests, and observations of actual contests demonstrate that the male that peaks at a lower roaring rate is most likely to withdraw at that point, or to lose the fight if the encounter escalates that far. As we noted earlier, the fighting ability (RHP) of an adversary may not be at all straightforward to assess; probing and monitoring feedback may be the only way to uncover the patterning of behavior indicative of different levels of fighting ability.

Rate of signaling is one of several contextual sources that are data driven (as this term is used above). Data-driven sources can arise either from the signaler or from the setting (Leger, 1993). From the signaler, the signaler's concomitant nonsignaling behavior is a useful source in addition to rate of signaling. Concomitant behavior can include spatial factors, such as how close the signaler is and whether it is oriented toward the target, factors that can influence the target's behavior more than signaling does (Paton, 1986). Sources from the setting include the behavior and identity of other nearby individuals, both heterospecific and conspecific. In field playback experiments, for example, white-throated sparrows respond more strongly to playbacks of a stranger's song than they do to playbacks of their own song. But this difference seems to be the result of the reactions of the playback subject's neighbors. The neighbors responded more strongly to the stranger's song than to the subject's (their neighbor's) song, approaching aggressively at the sound of a stranger (Brooks & Falls, 1975; Leger, 1993).

The assessing individual has also been identified as a source of contextual influence on the significance of signals (Leger, 1993). In the terminology used above, these are called state-driven contextual sources, a category of source that follows logically from the pragmatic approach to assessment developed here. That is, the significance of a perceived signal depends on the assessing individual's dominant regulatory problems at that time. Such influences have already been mentioned, e.g., the effect of nutritional state on Belding's ground squirrel responses to playback of antipredator vocalizations. Many of the identified sources of contextual influence on signal assessment undoubtedly also influence the individual's regulatory problems. These sources include the following.

1. *Stage of development.* For example, adult and juvenile California ground squirrels differ in how they differentiate among antipredator calls. Whistles are associated with high-urgency situations, and are most often elicited by raptors. Chatters are associated with less urgency, and are typically emitted while dealing with terrestrial predators, such as coyotes and bobcats. The reactions of adults to playbacks are consistent with this contrast in urgency. They spend more time out in the open and vigilant after chatters, and more time out of view and under cover after whistles. Juveniles reverse this difference, opting for more exposed vigilance after whistles, and more use of cover after chatters (Hanson, 1995). Developmental changes in the means and ends of regulation are discussed in Chapter 1, and are the topic of a later section of this chapter.

2. *Physiological (including motivational/emotional) condition.* For example, the regulatory problems of acquiring sufficient nutrition and avoiding predators can conflict with each other, in part because dropping the head to feed is incompatible with elevating the head to scan for predators. As discussed earlier, Belding's ground squirrels differ in the priority that they allocate to avoiding predators, depending on whether they are ahead of or behind schedule on weight gain in preparation for hibernation. When behind schedule, they take less time away from feeding to remain vigilant after playback of an antipredator call (see above).

3. *Previous experience.* For example, when vervet monkeys are exposed repeatedly to playbacks of a particular type of call by a particular individual, they become less responsive to such calls, but only to those from that individual. This suggests that they have come to interpret that individual as an unreliable source of cues (Cheney & Seyfarth, 1988).

4. *Relationship with the signaler.* For example, maternal northern fur seals alternate between two- to three-day nursing sessions and five- to twelve-day foraging trips during the four-month nursing period (Insley, 1996). These periods of absence from the pup are the longest of any pinniped, so each of the mother's returns is after an extended absence and to a crowded onshore colony containing many youngsters eager to suckle. Behavioral observations indicate that the effort to reunite is a mutual one. The mother begins calling as she comes ashore, and responds quite aggressively to suckling attempts by pups that are not her own. Her pup returns her calls with vocalizations of its own, and their reunions are remarkably quick, given the crowded, noisy conditions. Vocal playback studies demonstrate that mother–offspring recognition is mutual; both mother and pup vocalize significantly more to each other's calls than to those of other individuals in the colony.

4.4.2 Memory

When we speak of memory, we typically think of individuals retrieving information stored as a result of experience. However, animals also 'remember' things that they have never experienced (Coss, 1991). For example, laboratory-born California ground squirrels which have never experienced a snake before, seem to 'remember' what to do about snakes on first contact, tail flagging and substrate throwing at the snake, as well as apparently inferring that a snake who has been removed has crawled into a burrow (Owings & Coss, 1991). Such memories may persist for as long as 300 000 years in the absence of selection arising from snake predation (Coss, 1991). Qualitatively, this antisnake behavior is much like that exhibited by an experienced adult California ground squirrel. Similarly, the swamp sparrows and song sparrows discussed under *perception* above 'remember' some of the details of what conspecific song sounds like without prior experience with song (Marler & Peters, 1989). In both cases, these memories can contribute to the significance of vocalizations to assessing individuals. And, in both cases, these behavioral systems do change as individuals mature, no doubt in part because of memory changes arising from the squirrels' experiences with snakes (see also Hersek & Owings, 1994), and the sparrows' experiences with conspecifics. The storage mechanisms and retrieval processes for these innate memories may be basically similar to those acquired through experience (Coss, 1991).

Memory can stand an individual in good stead during agonistic encounters. During aggressive interactions, the owners of territories have an advantage over intruders; owners typically win many more of these encounters than intruders do. It is widely believed that familiarity with the territory lies at the root of this advantage. Familiarity could make the territory more valuable to the owner than to the intruder because the owner would have more knowledge of the locations of resources. This greater value should provide greater motivation to the owner than to the intruder during confrontations. However, Stamps (1995) emphasizes that familiarity has another effect, too; owners may *use* a territory more efficiently than intruders during an encounter, and so fight more effectively. Most of the thinking about resident advantage has focused on information (e.g., knowing the layout, knowing where the resources are, etc.) But, Stamps' motor-learning hypothesis focuses more on pragmatic issues of using a territory, rather than information, and in this sense is consistent with the pragmatic approach prescribed by A/M.

The distinction that Stamps makes, between memory of the layout of an area and memory of how to use an area, parallels in many ways the distinction in the cognition literature between semantic and habit memory (Squire, 1994). Research with both humans and nonhumans indicates that similar distinctions between types of memory systems apply, and that these multiple forms of memory are supported by different brain systems and have different characteristics. Habit memories are about what to do about a situation, and are similar in many ways to what Jean Piaget has called sensorimotor intelligence (Piaget, 1971). Semantic memories, on the other hand, involve more detailed cognitive representations of the situation itself. These two differ in their flexibility; semantic memories can often be linked to more than one form of action, whereas habit memories are more tightly linked to specific ways of behaving. For example, when a person learns a motor skill, such as a complex dance or the swing of a squash racquet, its form is often not consciously accessible (e.g., for verbal description), but only expressible through performance of the movements.

We suggest that much of the action in animal communication happens via processes much closer to habit than to semantic memory. This suggestion is consistent, for example, with the primary conclusions of an extensive program of research on the cognitive abilities underlying vocal communication by vervet monkeys (Cheney & Seyfarth, 1990). Vervet monkeys reveal considerable cognitive sophistication in their social maneuvering, but much of that sophistication is specific to the social domain, and not readily transferable to other domains, such as dealing with predators. Consistent with the pragmatic perspective of an A/M approach, the knowledge of assessment systems is pragmatic knowledge, about the implications of input for managerial action. Indeed, the process of developing social competence has been likened to a dance between the developing individual and its social companions (Kraemer, 1992). This topic is further explored later in this chapter with discussion of ontogeny.

4.4.3 *Referential specificity in animal signals*

From an informational perspective, a common way to think of signals is as 'standing for' or 'referring to' something else. It is primarily from this perspective that interest has originated in the possibility of referentially specific communication. A signal is deemed referentially specific when it conveys 'sufficient information about an event for receivers to select appropriate responses' without the aid of contextual information (Macedonia & Evans, 1993). Confident conclusions that a signal refers to a specific environmental event should be founded on study of both the management and assessment sides of the communicative process; that is, the relationship between eliciting events and signal structure must be described systematically (management side), and then the effects of variation in signal structure on target response must be studied through the use of playbacks (assessment side) (Evans, 1997). For example, one would need to demonstrate that the differing calls typically used by ring-tailed lemurs for aerial and terrestrial predators do not change with such factors as the proximity of the predators (Pereira & Macedonia, 1991), and then seek differences in response to these two classes of vocalizations in playback studies (Macedonia, 1990).

The possibility of high referential specificity in animal signals first achieved visibility because it was thought to indicate that animal cognitive abilities are more sophisticated than previously thought. Peter Marler and his colleagues initiated this inquiry (Marler, 1984). The ability to engage in referential communication was said to indicate that animals are more than simple, mechanical entities. Instead, referential signaling seemed to suggest that animal behavior can be mediated by quite specific cognitive representations, and not just by diffuse internal states. Interest in referential animal communication is just a small part of the 'cognitive revolution' that has recently transformed the study of animal behavior (Dyer, 1994).

More recently, referential communication has become a topic of interest in its own right, and researchers have become more cautious about detailed inferences of cognitive sophistication based on such work (Marler, Evans & Hauser, 1992; Macedonia & Evans, 1993; Evans, 1997).

How is referential specificity treated in an A/M approach? First, an attempt would be made to link the idea to fundamental systems of animal behavior, avoiding excitement about a phenomenon simply because it is relevant to apparently special human abilities. In that spirit, we suggest that the ability to treat perceived stimuli as standing for relatively specific environmental events is widespread, as is indicated by the fact that so many species are capable of the form of learning that is induced by Pavlovian conditioning procedures. Pavlov's dogs, for example, behaved as though the sound of a buzzer stood for meat powder after the two were paired repeatedly. But it would also be necessary to be precise about what is meant by saying that a predictive event (e.g., a conditioned stimulus, or vocal signal) 'stands for' a predicted event (e.g., an unconditioned stimulus, or predatory attack). In neither natural signaling systems nor responses to Pavlovian conditioned stimuli do animals treat the predictive and predicted inputs as equivalent. They behave as though they anticipate the arrival of the predicted event, not as though it has arrived. Responses to playbacks of antipredator signals rarely match the intensity or details of the form of actual reactions to predators. In most cases, for example, responses to calls involve silent vigilance; the vigilant responder engages in vocalizing and other evasive activities only if it detects the predator itself (e.g., Leger & Owings, 1978; Cheney & Seyfarth, 1990; Weary & Kramer, 1995). Similarly, responses to conditioned and unconditioned stimuli are typically not identical (Fanselow, 1991)

In an A/M approach, the contrast between the processes of management and assessment in communication is fundamental in thinking about referential communication. Even though communication involves an interplay between these two processes, the specific mechanisms involved are not necessarily intricately coordinated or coevolved. This point follows logically from the idea that assessment systems are specialized for many purposes other than communication, and so may bring pre-existing biases to a communicative context that shape the features of signals through feedback processes (e.g., Ryan, 1994). Nevertheless, the literature on referential specificity has, so far, assumed symmetry of the relevant mechanisms. While the need to examine both sides of the process has been emphasized (Evans, 1997), a referentially specific signal has been assumed, at least implicitly, to involve referential specificity in both the management system and associated assessment mechanisms of typical targets. Little consideration has been given to the possibility that a management system might meet the criteria for referential specificity, but not the associated assessment system, or vice versa (Owings, 1994). But this is probably a common mix. For example, we know from Cheney and Seyfarth's (1990) work on vervet monkeys that manager and assessor systems can function at different levels even within the same individual. The reactions of six- to seven-month-old vervets to call playbacks are indistinguishable from the responses of adults, but the emission of calls in adult-like fashion, confining each to a particular type of predator, does not develop fully for another 18 months. So, there is A/M symmetry when these juveniles respond to adult calls, but not in the equally likely case of adults responding to juvenile calls. Such asymmetry may be common in communication not only between different age classes, but also between different species.

An additional point is important regarding referential specificity from an A/M approach. The term *referential* specificity would be modified, to the less interpretive one of *situational* specificity. From the perspective of management, signals do not *refer* to anything; they are pragmatic acts emitted to produce an effect of variable specificity. This would lead us to treat the idea of referential specificity as important primarily as a subset of the broader topic of the relationship between signal structure and situations of use. As previously noted, according to A/M, relations between signals and situations arise from the shaping effects of situational constraints. A minimum requirement for situationally specific signals of the 'referentially specific' sort might be the possession by signal targets of the cognitive ability to deal with specific categories of entities, such as 'aerial predator,' and 'terrestrial predator' (Fig. 4.11). Such a cognitive ability would set the stage for managing the behavior of others through the use of signals with that level of situational specificity. Thus, the cognitive abilities of targets of management comprise one class of constraints on the specificity of the relationship between signal structure and situation. This class of constraints is the one most compatible with an

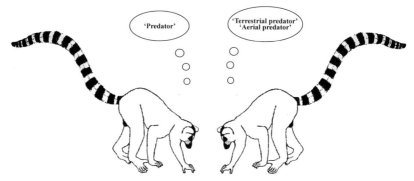

Fig. 4.11. Constraints of the cognitive categories of assessment systems on the potential predator specificity of antipredator vocalizations. These are ring-tailed lemurs. The assessment system on the left (a single, undifferentiated cognitive category for predators) would not select for different calls to aerial and terrestrial predators, but the assessment system on the right would.

informational view of signaling, i.e., 'if my conspecifics can understand about terrestrial predators, then I can signal to them about terrestrial predators.'

The above cognitive source of situational specificity offers little guidance regarding the expected structure of signals. (Chris Evans (in press) has converged with our thinking on these matters.) Macedonia and Evans (1993) have discussed sources of selection for contrasts in signal structure in an essay on the meaning of mammalian antipredator vocalizations. They begin with two simplifying steps, to make their task more tractable. First, they set aside the difficult question of underlying cognitive mechanisms, and explore 'functionally referential' signals, i.e., signals that are confined to specific predators and evoke predator-specific responses, whatever the associated cognitive mechanism. Second, they argue that the judgment of functional referentiality is least ambiguous when dealing with qualitatively different signals, rather than with different variants of the same signal type. (A signal that exhibits graded variation might be more likely to convey information about graded changes in motivational states.) Such qualitative differences in the structure of antipredator signals, they propose, might be favored by selection where the escape responses required to avoid the different classes of predators are incompatible with each other (e.g., vervet monkeys climb trees to evade leopards, but leave trees to deal with eagles). The rationale for this hypothesis is not entirely clear. Perhaps they mean that telling another individual to engage in conflicting activities might most effectively be accomplished with signals of conflicting structures; the contrast in structure would then supplement the contrast in meaning.

An A/M approach would also lead us to seek more pragmatic constraints. Acredolo and Goodwyn's (1990) work on communicative gesturing in normal human infants provides an example of this sort of constraint. Prior to developing verbal proficiency, human infants not only use nonverbal vocal signals such as crying and laughing; they also develop and deploy communicative gestures in their interactions with parents. The form of gestures for objects typically arises either within interactive routines involving the parent (e.g., mimicking the mother's puffing sound of blowing on a fish mobile as a general symbol for fish), or through the child's own interaction with the object (e.g., a throwing motion to symbolize ball). Thus, the forms of these signals are not arbitrary; they emerge from the interactions that the infant typically has in association with the objects. This proximate process of creating signals is analogous to the ultimate process of evolving a signal from an intention movement (Tinbergen, 1952; Smith, 1977). Regarding the origins of form differences in antipredator calls, these results would lead us to explore how, for example, different evasive maneuvers might differentially distort the vocal cavity during calling; that is, how call structure emerges from interactive routines with the predator. Ohala (1984) has made an analogous 'byproduct' argument regarding the origins of the appeasing 'grin' and aggressive 'o-face' in primates (see Chapter 1). These visual displays, he hypothesized, were originally simply the facial adjustments needed to raise voice pitch in appeasement (shortening the vocal tract with the grin), and to lower voice pitch in a threatening growl (lengthening the vocal tract via lip protrusion in the 'o-face'). These facial expressions subsequently became emancipated from sound production, serving as signals in their own right.

Finally, M/A's focus on the broad issue of constraints can direct our attention to sources of situational specificity that would not appropriately be called referential even in an informational perspective. (Again, see also Chris Evans' convergent proposals, Evans, 1997). For example, Macedonia and Evans' hypothesis, described above, would lead to the prediction that California ground squirrels should have what they call functionally referential antipredator signals. Incompatible antipredator responses are required for two types of predator; the burrows in which these squirrels live are sources of refuge from most mammalian predators that threaten them (not counting mustelids), but are sources of danger from rattlesnakes. Consistent with this hypothesis, the squirrels use very different signals for the different types of predators, chatter calling for mammalian predators, but waving their fluffed tail back and forth (tail flagging) for snakes (Owings *et al.*, 1986). By their operational definition, then, this is a functionally referential signaling system. However, the term seems inappropriate, for the following reason. The most likely reason for the structural difference in signals is that the predators themselves are targets of these signals (Hennessy & Owings, 1988; Hersek & Owings, 1993). The switch to visual signals with snakes apparently has been favored because, unlike mammalian predators, the snakes apparently cannot hear the calls of these squirrels. The specificity of signal structure to predator type arises from the properties of the signal target, not what is typically called the referent. Extending Macedonia and Evans' argument, in our terminology, we might say that qualitative differences in constraint on signal structure arise from a variety of sources, which can, for example, include the sensory specializations of signal targets.

4.4.4 Deception

On the day that a female Formosan tree squirrel becomes sexually receptive, 9–17 males assemble near her home range to attempt to mate with her (Tamura, 1995). During her approximately eight hours of sexual activity that day, she consorts with about eight males in succession, copulating once or several times with each. After mating with the female, most males initiate a series of 'postcopulatory calls,' loudly barking repeatedly for about 17 minutes. Other squirrels in the vicinity respond to these calls with apparent alarm, retreating from the ground to tree canopies and remaining immobile there for the duration of the calling bout. By calling, the consorting male may delay the next male's access to the female, which may increase the proportion of the female's litter that the consorting male sires. It is known that in some squirrels the proportion of a litter that a male actually sires depends in part on how quickly the female copulates with the next male (Schwagmeyer & Foltz, 1990).

Why do other squirrels seem to respond with alarm to these postcopulatory calls? Perhaps it is because they sound like the pattern of barking used when mammalian predators are detected. Indeed, statistical comparisons of the form and patterning of these calls revealed no significant differences between those emitted after copulation and those evoked by feral cats. And, playback experiments found that recordings of calling in the two contexts did not differ significantly in their effectiveness at inducing flight to trees and immobility.

Are these males deceiving other squirrels by using alarm calls to achieve their own selfish ends? We suggest that this is not the most scientifically valid question to pose. Similarly, we would argue that it is a misdirection of scientific inquiry to ask whether animals are *really* managing the behavior of others. In both cases, a more legitimate approach would be to ask how useful the concepts of deception and management are for making sense of the communicative patterns that we discover. This whole book explores the utility of the concept of management (and assessment). So, the focus here is on how useful the idea of deception is.

The considerable recent interest in the idea of animal deception (Hauser, 1996) has two sources: (1) recognition that natural selection can lead to exploitative behavior toward conspecifics (Burghardt, 1970; Dawkins & Krebs, 1978); and (2) an anthropocentric emphasis of concepts created to describe human behavior (Owings & Morton, 1997). We concur with source one, but resist anthropocentrism. The concept of deception is anthropocentric, in that it has been applied primarily to human interactions. This anthropocentrism is compounded when deception is defined as conveying false information or withholding true information. The primary goal here is to explore deception without appealing to information exchange. The concept of deception will be critically evaluated, retaining it but applying it in a more restricted way than is typical.

What are the phenomena that we often call deception? A mismatch occurs between our perception of how an individual *should* be behaving with regard to communication, and how it *is* behaving. The Formosan tree squirrels described above provide an example. Once we know the kind of calling that is used to deal with mammalian predators such as cats, we are surprised to discover the same kind of calling in the very different context of the aftermath of copulation. Such deceptive mismatches between our expectations about an animal's communicative behavior and the individual's actual behavior can take two general forms. The animal may emit a signal that we do not expect in the observed context, or fail to signal where we expect it. In informational terms, the former is said to involve the provision of false information, and the latter is labeled withholding information.

The decision that the animal is behaving deceptively is an onerous one. That is, we need to evaluate critically our assumptions about what the function of the signal is; this can be done in part by acquiring solid data on the norms of use of the signal before concluding that the norms have been violated (Smith, 1986a; Smith, 1997). Such steps are not always taken. Male barn swallows, for example, have been reported to use anti-predator calls deceptively as a means of disrupting copulation between the caller's mate and another male (Moller, 1990). But, no quantitative normative data are available that would permit us to judge whether these calls really do function primarily as an antipredator warning. A much richer data set is available for Formosan tree squirrels; the distinctive pattern of repetitive calling is the norm both during encounters with cats, and in the aftermath of copulating. Such data complicate efforts to judge which context of calling is the original or normative one. Logic seems the primary guide; it is clearer how males would benefit from using antipredator calls after copulating, than how they gain from using post-copulatory calls to deal with predators. And, the cautious reactions of squirrels to playbacks of these calls seem more appropriate for an anti-predator context than for a mating one.

Foregoing signaling has been proposed to be a safer way to behave exploitatively than emitting a signal in an atypical situation, because foregoing leaves the individual less conspicuous and therefore reduces its chances of being 'caught' (Cheney & Seyfarth, 1985). According to this hypothesis, foregoing calling should be more prevalent and therefore easier for biologists to discover. The food-associated calls of rhesus monkeys may provide an example; they often call when they discover food, but are also frequently silent (Hauser & Marler, 1993b). When noncallers are discovered with food, their chances of being the target of aggression are greater than if they had called. The absence of food calling has been described as a case of deception about possession of food, and the increased incidence of aggression where calling did not occur has been interpreted as punishment for cheating (Hauser, 1992). Although this is an intriguing example, the case that not calling is deceptive will require additional scrutiny. For example, calling becomes more likely when the discoverer of food is hungry; so, a noncaller may simply be sated. Also, not calling was actually more common (55 percent of trials) than calling. It would normally be expected, from models of Batesian mimicry, that cheating would be effective primarily when rare; if cheating becomes the norm, the associated signal should lose its capacity to influence the behavior of others (Harper, 1991).

Hauser (1997) has made headway in building a case that not calling is deceptive, in part by distinguishing among different sources of the absence of calling. He has discovered that not all cases of failure to food call are appropriately labeled as deception; during experimental exposures to food, peripheral males (in contrast to regular members of troops) never called. Since calling is not the norm, then not calling could hardly be considered deceptive. And, when other monkeys discovered peripheral males with food, they did not behave aggressively toward them. It is primarily calling by troop members that is the norm, and predominantly these members who experience elevated aggression when they do not call. This pattern makes sense; punishing peripheral males may not be useful, because the punisher may not encounter the victim again, and so may not benefit from the effects. But, punishing a member of the same troop could generate benefits that exceed costs, because of the high probability of future contact between punisher and punishee (Clutton-Brock & Parker, 1995) .

The rhesus monkey example draws us back to several themes that are central to this book: the importance of distinguishing the perspectives of management and assessment, and the critical role of assessment in communicative processes. The idea of deception seems less useful from the perspective of management, given the pragmatic emphasis of this perspective. An individual that foregoes calling where it is the norm to do so may simply be protecting itself from the self-interested actions of others, in a way similar to the efforts of a camouflaged animal to limit its conspicuousness to a predator (Mitchell, 1986). Indeed, food calling by rhesus monkeys does attract others who are likely to take some of the food, and individuals get to eat more when they forego calling and go undetected than under any other conditions (Hauser & Marler, 1993b). When discussed in this way, foregoing calling does not seem like 'cheating;' where are the 'rules' being violated by such behavior?

An interpretation in terms of camouflage has also been applied to 'deceptive' efforts by male pied flycatchers to mate bigamously (Dale & Slagsvold, 1994). Males of this European species of songbird sing to attract their mates to the nest cavities that they defend. After mating with one female and producing a clutch of eggs with her, they travel to a nest hole a considerable distance from their current territory (an average of about 200 m) and attempt to attract a second female. When they are successful, they produce a second clutch of eggs with her, but then leave her on her own, returning to their primary female to assist in rearing those young. The reproductive success of the primary mates of bigamous males is close to that of monogamously mated females, but the deserted, secondary females are significantly less successful (Alatalo *et al.*, 1981). Are these males deceiving their second mates, establishing a distant second territory to attract a second mate by hiding the fact that they are already mated? Perhaps not; maybe their move to a distant territory is a pragmatic tactic to minimize disruption of the second mating effort by the first mate (Dale & Slagsvold, 1994). As we have argued from an A/M perspective, the best way to understand why animals use the signals they do, and in the ways they do, is to appeal to the constraints of the signaling situation, such as the impact of first mates on efforts to acquire second ones. This example, like the case of rhesus monkeys who forego food calling, highlights the limitations of anthropocentric terms like *deception* and *cheating* when applied to the behavior of nonhumans. The concept of cheating especially implies the violation of a set of rules or laws. Is it useful to think of nonhumans as guided by rules of conduct like the rules and laws of human cultures? We suggest that such concepts can distract us from exploring the constraints of communicative situations, generating controversy instead over whether animals are *really* being deceptive.

Dale and Slagsvold (1994) extend their thinking about pied flycatchers in ways that are also consistent with an A/M approach, noting the importance of considering the question of deception from the perspective of the assessing females. Their treatment of the females' assessment processes is consistent with signal-detection theory (Getty, 1996), a model of decision-making under uncertainty that psychologists adopted from engineering several decades ago (Tanner & Swets, 1954). The general idea of signal detection theory is illustrated in Figure 4.12. Male pied flycatchers vary in the amount of time they spend singing on their territories, but unmated males spend more time on average singing on their primary territories than mated males do on their secondary territories (Dale & Slagsvold, 1994). The uncertainty that a female faces in discriminating between mated and unmated males is indicated by the overlap in these two hypothetical distributions. Because of this overlap, the female's decision as to whether to mate with a particular male or not becomes a statistical matter; that is, she must always deal with the possibility of making a 'mistake,' either by rejecting an unmated male, or by accepting an already-mated male. Notice that, for a given level of difference between mated and unmated males, reduction in one type of error leads to an increase in the other. So, where her criterion for mating should be placed depends on the relative importance of the two types of errors. For example, in the short breeding season available to these birds in northern Europe, rejecting an unmated male could be fairly serious because another might not come available until it is too late to complete a cycle of reproduction. So, the female might be expected to set her acceptance criterion relatively low, at point *a* rather than *b*, to reduce her risk of

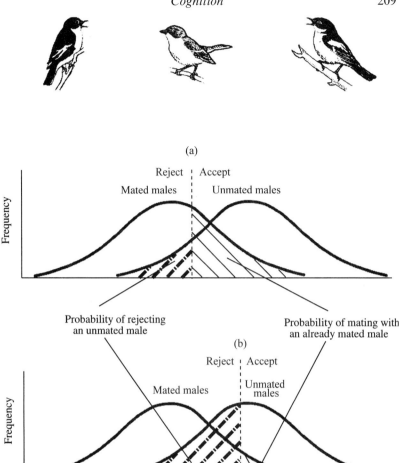

Fig. 4.12. A signal-detection theory approach to mate choice by female pied flycatchers. Females reproduce more successfully as primary or only mates. So, females face the task of choosing unmated males over males on a secondary territory who are already mated with a female on their distant primary territory. Mated males are known to be potentially distinguishable, because they spend less time on average singing on their secondary territories (left-hand distributions) than unmated males do on their only territories (right-hand distributions). But, discriminating the two categories of males involves uncertainty in proportion to the degree of overlap of the two distributions. Decision criterion (a), at the top, minimizes the mistake of rejecting an unmated male, but at the cost of a greater risk of mating with an already-mated male. Decision criterion (b), at the bottom, minimizes the mistake of accepting an already-mated male, but at the cost of a greater risk of rejecting unmated males. See text for additional details.

rejecting an unmated male. Notice the bind she is in, though; by lowering her criterion, she has increased her risk of mating with an already-mated male. Such constraints on potential mate assessment by females probably account as much for 'deception' in pied flycatchers as devious managerial activity by males does.

Deception: proximate contexts and mechanisms

It was noted earlier that the idea of deception is typically applied when a discrepancy is discovered between what animals should be doing, in our opinion, and what they are doing. However, not all such mismatches are considered deception. Some have been labeled mistakes, especially when the individual is young. An infant vervet monkey, for example, is said to be mistaken, not lying, when it responds to a warthog by barking, a call that adults generally use only for dangerous mammalian predators such as leopards (Cheney & Seyfarth, 1990). Reservations about treating such actions by infants simply as mistakes were discussed in Sections 1.2.9–1.2.11, and it was suggested instead that infants might be using barks as a means of evoking immediate parental feedback and assistance, if needed. This point highlights the importance of critically evaluating our assumptions about the functional organization of the individual's behavior. Youngsters are not 'trying' to act like adults, but falling short; they are behaving under the distinctive constraints of their current developmental stage. And, they are not applying correct versus incorrect, or true versus false labels to events; they are taking care of themselves, in part by making use of resources extractable from the parent. More generally, signals are not most usefully thought of as statements of fact that can be judged to be true or false; signals are more like the human speech acts discussed below – outputs that serve to achieve some effect on targets.

The vervet monkey developmental example above highlights the importance of understanding the mechanisms and proximate contexts of communication. That is, when one understands that young animals use signals appropriately for their ontogenetic niche, one is less disposed to treat them as mistakenly applying vocal labels to the wrong animals. Similarly, when one understands the proximate constraints that male and female pied flycatchers work under, the concept of deception becomes less useful. The knowledge structures that proximately underlie animal communication are discussed in Section 4.4.2, and it is proposed that they are much more similar to what has been called procedural memory than to semantic memory. That is, the evidence of domain specificity even in the relatively sophisticated cognitive systems of such species as vervet monkeys (Cheney & Seyfarth, 1990) suggests that their knowledge is tied to particular action systems, rather than being applicable more broadly as some aspects of human knowledge are. This indicates that animal knowledge structures are fundamentally pragmatic, i.e., about what *to do* about objects, events, and states, a finding consistent with the stance specified by an A/M approach. According to this approach, signals are not statements of fact, that can be judged to be true or false, but are efforts to produce certain effects.

Even human language, that most semantic of communication systems, has this pragmatic quality, according to the *speech-act* literature (Horn, in press). Many human speech acts are not true or false statements of fact, but efforts instead to accomplish specific goals. The following examples illustrate this point.

1. 'Get out!'
2. 'Take your hands off of me!'
3. 'I now pronounce you husband and wife.'
4. 'Please make yourself comfortable and tell me what's on your mind.'
5. 'If an emergency arises, line up single file and walk quietly and quickly to the shelter.'

Those who offer analogies to human language in animal communication should attend to the speech-act literature at least as much as to the literature on cognitive representation.

A dedicated proponent or devil's advocate of information exchange might respond that even the above examples of speech acts need to correlate with conditions important to assessing target individuals in order to be effective. Statement 3, for example, is valid only if issued by a Justice of the Peace, or ordained minister. If issued by a person on the street, the man and woman do not legally become husband and wife. Similarly, statement 5 will continue to be effective only if marching quietly in line to shelters proves to be an effective way to deal with emergencies. Are these correlates, an information advocate would ask, not the information made available by these signals? No; this question confuses time frames of causation. Such correlates *validate* the cues as useful to assessing individuals (Horn, 1997), but are not the immediate cause of the assessing individual's reaction to the statement. Note that Horn's proposal turns the usual view of the role of information on its head. The correlates of signals ('information') are not immediate causes of the behavior of targets of signals, they are instead long-term validators of the signal's utility.

4.5 Development

According to an A/M approach, individuals are complete A/M systems within their natural developmental niche even at birth. In other words, their behavior must be understood in terms of how it meets their needs now, and not just how it contributes to maturation. These complete individuals are born into an environment consisting of other animals that are at least as complex and actively self-interested as the developing individual. So, from the outset, effective management necessitates negotiation and even competition with companions.

These managing/assessing companions are often very reliable sources of feedback about which managerial tactics are most effective, and under what circumstances. This feedback does not necessarily arise from teaching; it is often a product of the other individuals' pursuit of their own interests. Without this daily give and take of life with companions, individuals develop only limited effectiveness, at best, in managing the behavior of others (Mason, 1979c; West, King & Freeberg, 1997). Such developmental effects of companions can be illustrated by West, King and Freeberg's work on the ontogeny of competence as a singer in male brown-headed cowbirds. If young males are reared with the opportunity to interact only with older conspecific females, but not with older males, they develop songs that are actually *more* effective than those of normal males at evoking copulatory postures in females during playback studies. How could 'deprived' males develop more potent songs? Doesn't this contradict the statement above that social feedback is essential for the development of social competence? Only apparently. It is important to note first that these young males were not deprived of feedback from females; the selective responses of females to particular song structures did in fact play a role in shaping the potent songs that the young males came to sing (West, King & Freeberg, 1994). But, the absence of feedback from older males had a profound effect on the competence of these young males. Even though they had very potent songs, they did not deploy them effectively. Normal males sing as part of energetic efforts to court females, approaching them, maneuvering to get in front of them, fluffing their feathers as they get close, and mounting when the female invites it. In contrast, the male-deprived males did sing at females and even occasionally mounted them, but they lacked the normal males' enthusiasm and focus, quickly lapsing into a state of disinterest, perching, and 'vocalizing in a manner closer to hiccuping than interacting.' Despite their more potent songs, deprived males achieved many fewer consortships than normal males did.

Such results indicate that the development of individuals is channeled by two inherited sources of influence. When we think of inheritance, we usually think of inherited genetic programs. However, genetic programs are selected to induce only those developmental outcomes that are not reliably generated by the environment. Developing individuals also inherit the environments with which they must negotiate (West *et al.*, 1994).

At any age, other organisms play key roles in an individual's fundamental regulatory processes. These other organisms may be sources of food or competitors for it, cues about the presence of predators, refuge from predators or sources of predatory threat, warmth to huddle with in the cold, collaboration in bringing up offspring or infanticidal threat to young, and so forth. Such contributions by one individual to the regulatory processes of another are perhaps most conspicuous in parent–offspring relationships. For example, the role that the mother plays in a rat pup's thermoregulation has already been discussed. The mother's huddling with and retrieval of her pups plays a critical role in the pups' maintenance of body temperature. And, of course, the pups orchestrate such maternal contributions with their ultrasonic vocalizations. It is only at later ages and larger sizes that pups become more self-sufficient in their thermoregulatory abilities (Ryan & Rand, 1993b).

A conspecific constitutes a package of influences on a developing individual (Lorenz, 1970). Discovery of the contents of this package has required much careful experimentation. For example, a long period of separation of a rat pup from its mother (18 + hours) induces an initial period of persistent retrieval calling by the pups (see above), and a later phase involving a complex of other changes that seem superficially like the 'despair' exhibited by human infants during long-term separation. However, detailed analysis of the properties and sources of these longer-term changes illustrates the mother's many separate contributions to the pup's regulatory processes, and the inapplicability of the despair concept (Hofer, 1987) . The separately regulated processes in the pup include its gross motor activity level, sucking behavior, oscillation between sleeping and waking, and release of growth hormone. Different kinds of maternal stimulation produce increases and decreases in the level of these various processes. For example, activity level in a novel environment increases in pups isolated from their mothers but artificially maintained at appropriate nest temperatures. This increase begins approximately four hours after separation and reaches an asymptote after about 18 hours. It is not nutrient levels that normally limit activity levels; contact with the mother blocks this increase, for example, even when she cannot nurse her pups. It is the olfactory and tactile stimulation provided by the mother's periodic visits to the nest that tends to keep activity down to normal levels. On the other hand, if pups are not artificially kept warm during these long-term separations, their activity undergoes a decrease rather than an increase. This suggests that the thermal stimulation provided by the mother tends to increase activity levels, and that the opposing effects of thermal and tactile/olfactory stimulation tightly maintain pup activity levels within a narrow range.

Such maternal contributions to young mammals' management of their own situations does not cease once youngsters become more mobile and thermally self-sufficient. As four-week-old Belding's ground squirrels first begin to emerge from their nursery burrows, they face serious danger from predators (Mateo, 1996); 13–27 percent of them will disappear during their first two weeks above ground, primarily as a result of predation. So, they would benefit from being able to respond appropriately to evidence of a nearby predator. Playbacks of the antipredator calls of this species indicate that the pups rely on their mother initially, but quickly become more self-sufficient. On the first day of emergence, pups are more likely to respond to playbacks if their mother is present than if she is absent. When they do respond, they typically bolt into their burrow, leaving the job of surveying for predators to their mother. By day five, pups react to playbacks whether or not their mother is present, and they remain outside their burrow, standing tall on their hind legs as they survey for the source of alarm themselves. (This rapid development of self-sufficiency may be in preparation for independence only four to six weeks after emergence, at about ten weeks of age (Holekamp & Sherman, 1989).)

Infant vervet monkeys also appear to use adult guidance. If they look toward adults before responding to an antipredator call, their behavior is more likely to be appropriate for the call's typical predatory elicitor than if they do not glance at an adult first (Cheney & Seyfarth, 1990). There is evidence that infants actively solicit this adult guidance, rather than simply using it when available (Owings, 1994). Infant vervet monkeys emit antipredator calls at a much wider variety of disturbances than adults do (Cheney & Seyfarth, 1990). Calls by infants at harmless species have been treated as mistakes (Cheney & Seyfarth, 1990), but there is evidence in this same research program that calling infants receive immediate adult guidance and assistance (Owings, 1994). If the animal eliciting the infant's calling is a dangerous predator, then adults call, too; the infant, in turn, responds by fleeing to its mother, who protects it. If, on the other hand, the infant has called at a relatively harmless species, others are less likely to call, and the parent even occasionally punishes the infant (see also Caro & Hauser, 1992).

What are the developmental effects on the youngster of such maternal influences on regulatory processes (and paternal and peer, too, where there is regular contact with these types of individuals)? This question can be posed in two time frames – relatively immediate and longer term. With regard to relatively immediate effects, one can explore how these processes contribute to the individual's effectiveness as a manager/assessor while it is still young. With regard to the longer-term, one can ask how maternal regulatory processes lead to eventual development of adult levels of self-sufficiency.

4.5.1 *Immediate effects*

Relatively immediate effects can be illustrated with the copious research on filial imprinting in domestic chickens and their wild progenitors, Burmese red jungle fowl (for reviews see Bolhuis, 1991; Bolhuis & Van Kampen, 1992). Chick embryos first respond to sound at about 14–15 days gestational age (six to seven days before hatching). Two-way vocal interaction between the embryos and their mother begins on the day before the chicks hatch, and the rate of calling by both mother and offspring increases as the time of hatching approaches. When the embryo emits distress peeps, the hen calls or moves to the nest, and the embryo becomes silent or begins to emit pleasure notes.

During the first one to two days post-hatching, the chicks remain mostly in the dark under their mother, but very much in vocal contact with her. If they do become separated from their mother during this time, young chicks do approach and follow her and their siblings. Their readiness to follow is intensified by the *clucks clucks* that the mother emits repeatedly as she walks along with ruffled feathers. As a result of this association, the chicks develop a visually based attachment to these individuals, and the stimulation produced by the mother's *cluck clucking* enhances their subsequent attraction to her (Van Kampen & Bolhuis, 1993). Soon, when the chick is close to its mother and siblings, it will attempt to snuggle up, frequently emitting *whee* or *trill* vocalizations. If the chick becomes isolated from its family, it will switch to loud *peeps* as it searches to reunite, but will avoid approaching a strange hen.

Early on, the clutch of siblings sticks together tightly and rarely strays more than a meter from the mother. Such attraction of a chick to its family has great utility; the mother is a remarkable resource for the chicks during this time. With regard to their nourishment, for example, the chicks respond to their mother's food pecking by pecking at objects on the ground, and they learn to peck at the same kinds of food items that she does (Suboski & Bartashunas, 1984). The mother also uncovers food for young chicks, breaking food items up if they are too large, and passing them directly to her chicks. As they grow older and wander more distantly, chicks respond quickly to the mother's *duk duk duk* food calls, approaching and eating the food items that she points to with her beak (Stokes, 1971). Chicks also cope with predatory threats through their mother (Palleroni & Marler, in press). When alone, chicks have only one response to initial evidence of predatory threat: they freeze and go silent. In the presence of their mother, however, they adjust their behavior more appropriately for the form of predatory threat. Cueing on their mother or on a nearby cock, chicks scatter and take cover at the adult *screams* that aerial predators evoke. In contrast, the mother's *duk, duk, duk, kaaahs!* to terrestrial predators prompt the chicks to move away from cover and assemble about her, *peeping* repeatedly.

Social companions also have subtle and unexpected influences that fall outside the realm of what might be called 'guidance.' As noted briefly above, chicks begin distress *peeping* when socially isolated, and the arrival of either siblings or the mother is a source of feedback to the calling chick, reducing its calling. The chick's *peeping* is an effective means of helping to restore contact with its family; *peeping* increases maternal efforts such as calling to reunite with her young (Hughes, Hughes & Covalt-Dunning, 1982). In mallard ducklings, the tactile stimulation that comes from close association with the mother and siblings has a powerful calming effect that initiates a period of malleability, during which preferences develop for the vocalizations heard regularly during that time. In a natural context, these vocalizations are the maternal calls of the ducklings' mother, and the tactile stimulation comes from her and other ducklings. However, experimental work indicates that exposure to playbacks of heterospecific maternal calls, of a domestic chicken rather than a mallard, can induce a preference for the chicken calls as long as the requisite tactile stimulation is present. In addition, this tactile stimulation can be provided mechanically rather than by other ducks (Gottlieb,

1993). Thus, some kinds of stimulation come as parts of the social-companion package, and can contribute to learning during development without providing specific cues about what is to be learned. This process of creating a state conducive to learning, in contrast to processes of (incidentally) 'instructing' the developing individual, provides an example of motivational effects on the ontogeny of assessment systems.

The above discussion indicates that management and assessment processes within the same individual become linked to each other through their reciprocal influences on each other during development. It has been seen that tactile stimulation from family members sets the stage for development of such fundamental assessment abilities as recognizing conspecifics through their vocalizations. But, it has also been noted that developing individuals are not simply passive recipients of such tactile stimulation. Contact with others comes about in part as a result of the developing individual's own efforts; distress calling by a youngster, for example, attracts family members into contact with it. So, individuals play a significant role in orchestrating their own development, but they do so in part through the social consequences of their managerial activities (West *et al.*, 1997). These social consequences include incidental 'instruction,' in the sense that specific patterns of stimulation are learned and subsequently influence the individual's behavior. But, motivational effects are also important, in the sense that social consequences can influence development and learning, but do not become part of what is learned.

The motivational effects of social stimulation seem to be especially linked to its contingent properties, that is, to the parts of social input that are responses to the individual's own behavior. Young Japanese quail chicks, for example, develop a preference for an object that they see during their period of imprinting (ten Cate, 1986). The strength of this preference depends on certain properties of the object. If the object moves rather than remaining stationary, the chick becomes more attached to it, and if the object's movement is produced by the chick's distress calling, the attachment is especially strong. So, the chicks are most likely to become attached to the individual who most consistently responds to their calls, and, in the natural context, that individual usually happens to be the chicks' mother or siblings.

4.5.2 *Longer-term effects*

As discussed above, the neural representations of companions acquired early in life play a relatively immediate role. They are used by the developing individual to recognize and maintain an attachment to specific adults and peers. These same stored images of companions can also play a role later in life in the development of preferences for sexual partners (Bischof, 1994). This process of sexual imprinting involves two stages – the acquisition of the neural representation of companions, and the consolidation of sexual preference, in which this recognition system becomes linked to the executive neural system for the production of sexual behavior (Bateson, 1987; Bischof, 1994). The idea of two stages may sound familiar; as discussed below and in Chapter 1, a songbird's development of the ability to sing species-typical song also involves a two-stage process in which an early stored representation of normal song becomes linked to the vocal output system through a process of feedback-based practise. Sexual imprinting and song learning share the additional feature that social interactions significantly influence the outcomes of those processes (ten Cate, 1994; Bischof, 1994; Clayton, 1994). The ontogenetic processes underlying sexual imprinting and the development of song will provide the core for our exploration of longer-term developmental processes.

Zebra finches are one of the few species in which both song learning and sexual imprinting have been studied in some detail (see Clayton, 1994). This is an Australian species that exhibits a prolonged breeding season each year, extending for as long as seven months (Zann, 1990). Young become nutritionally independent at about 35 days of age, shortly after which their parents begin to re-establish their nest for another brood. At this time, the young often maintain their association with their natal colony, but join in flocks with adults for foraging and drinking sorties outside the bounds of the colony. Unlike many species of songbird (Baptista & Gaunt, 1994), female zebra finches do not sing. Therefore, the following survey focuses on males, the sex emphasized in studies of song development and sexual imprinting.

The processes of maturation of song structure and sexual preference follow apparently independent schedules. Neural storage of the relevant representations occur during restricted periods of the individual's lifetime, but at different times for the two processes. For sexual imprinting, storage is completed for the most part during the period from 13 to 19 days of age, but may under some conditions be extended (Bischof, 1994). The storage phase for song learning hardly overlaps at all with this dominant period of storage for sexual imprinting, occurring as it does from about 35–65 days of age (Slater, Eales & Clayton, 1988). Both sexual preference and song structure crystalize as individuals become sexually mature, at about 80–100 days of age (Zann, 1990; Bischof, 1994).

4.5.3 *Sexual imprinting*

Evidence that adult sexual preferences in male zebra finches are founded upon early social learning comes from cross-fostering experiments, in which zebra finches were reared by Bengalese finches (Immelmann, 1969). When tested as adults, such cross-fostered males exhibited a preference for female Bengalese finches over females of their own species. This preference is founded primarily upon the learned visual features of the foster mother (Bischof, 1994). For example, males were foster reared by zebra finch parents. In some foster-parent pairs, the fathers were of normal plumage and the mothers were an all-white domesticated form; in other pairs, mothers were normally plumaged and males were white. The fostered males subsequently preferred a female with the same plumage as the foster mother's (Vos, 1994)

The early learning in sexual imprinting is influenced by the frequency of social interaction, including vocalization, between the developing individual and its foster parents (Bischof & Clayton, 1991). The male within each litter that begged and called for food most from its Bengalese finch foster parents and that was fed the most by them developed the strongest preference for Bengalese finch females when given a choice between these and female zebra finches. However, this latter study indicated that a second phase of social interaction, the first courtship, was important in stabilizing the males' sexual preference. These males were reared with their Bengalese finch foster parents until 40 days of age and then isolated until day 100. After that, half of the males were exposed to a Bengalese finch female for one week and a zebra finch female during the next week, and the other half received similar exposures but in the reverse order. These highly sexually ready males actively courted these females. Finally, all males were tested for their sexual preferences in a simultaneous choice between Bengalese and zebra finch females. Those males exposed first to Bengalese females after day 100 unanimously preferred Bengalese females in the preference tests, singing almost exclusively to these females of the same species with which they had been fostered earlier. In contrast, those males exposed first to zebra females exhibited much more variation in the preference tests, some preferring zebra females, others Bengalese females.

The importance of the first courtship after sexual maturity in the development of sexual preferences reflects the operation of a second stage in this process (Bischof, 1994). In the natural context, the species first courted would be the parental species, and the experience would simply consolidate the sexual preference. Two kinds of evidence indicate that this second phase is one of consolidation rather than additional learning. First, other males in the above study (Bischof & Clayton, 1991) were reared by zebra rather than Bengalese finches. In contrast to the males fostered with Bengalese finches, these males exhibited a preference for zebra females irrespective of which species they were exposed to first after day 100. Second, actual exposure to a female need not occur at this time in order for consolidation of the sexual preference to take place. If a male's sexual motivational system is simply aroused, e.g., through exposure to a nest box and nest, this is sufficient to stabilize his sexual preference for the parental species, whether they were heterospecific or conspecific (Oetting, Prove & Bischof, 1995).

Bischof has hypothesized that this consolidation process does not involve the storage of additional details about the perceptual features of preferred mates. Consolidation is instead the establishment of connexions between the recognition system in which preferred perceptual features are stored, and the executive neural system that regulates the expression of reproductive behavior (Bateson, 1987; Bischof, 1994). In other words, the effects of consolidation are more motivational than cognitive. However, the experimental creation of a mismatch between the features of the currently courted conspecific female, and those of the heterospecific foster mother can initiate the establishment of a second neural representation, of the features of this second type of female. In this situation, the interaction with this second type of female has the potential to consolidate links between the executive system and the representation of both the foster mother and the current female.

The extent to which the second representation 'captures' any input sites depends on the number of sites that the representation of the foster mother is able to dominate. There is reason to believe that the number of these input sites is limited, so that strong connexion by one representation can 'competitively exclude' another (Bateson, 1987). Consistent with this idea is the finding that the half of the males that begged from and were fed most frequently by their Bengalese foster mothers exhibited very little shift in preference to zebra females as a result of exposure to them (Oetting *et al.*, 1995). In other words, their foster-mother representations occupied most of the available connexions during consolidation. In contrast, the half of the males fed least frequently by their Bengalese foster mothers exhibited a significant shift on average toward a preference for zebra finch females after exposure to them. The less-fed males varied in their preference shift, depending on their motivational state at the time of exposure to the zebra female. The higher the male's arousal during exposure to the zebra finch, the stronger his subsequent preference for the zebra female. According to Bischof's hypothesis, higher male arousal while with the zebra finch female generated capture of a larger proportion of inputs by the representation for zebra females during consolidation, and therefore a stronger shift toward preference for the zebra female.

In summary, to communicate competently during courtship, a male zebra finch must be able to direct singing and other activities toward the appropriate species of female. This is important, since zebra finches are known to associate in mixed-species flocks during the breeding season (Zann, 1990). The developing male's own social initiatives generate many of the experiences necessary for development of this ability. Prior to fledging, the male's visual and vocal signaling during food begging stimulates feeding by the parents; these experiences yield a neural representation of the mother in the fledgling that can subsequently guide selection of a target for courtship. But the assembly of the courtship system into a functional mechanism depends on additional initiatives by the developing male. The first efforts at courtship generate connexions between the previously established recognition system and the neural systems that regulate expression of courtship behavior. This consolidates the mechanisms underlying male courtship behavior and sets the stage for effective deployment of courtship signals.

Of course, this summary represents only a small part of the whole developmental story. For example, some of the components of the courtship executive system are similarly dependent on the give-and-take of social interaction for their normal development. The ability to sing normal song is one example, which is the topic of the next section.

4.5.4 Song learning

When Immelmann (1969) cross-fostered zebra finches with Bengalese finch parents, the zebra finches not only preferred Bengalese over zebra females, but also sang songs that resembled those of Bengalese finches. The processes involved in development of song have since been explored in some detail.

As noted above, young zebra finches become nutritionally independent of their parents at about 35 days old. They form small, inconspicuous groups within the colony at this time, but begin to flock with other adults around day 40 for feeding and drinking forays away from the colony. While in the nesting colony, immature birds remain relatively inconspicuous, but males begin their first efforts to sing around fledging time, and solitary males can be heard singing loud bouts of subsong from the tops of bushes (Zann, 1990). If we can extrapolate from observations of other species (West & King, 1985; DeWolfe *et al.*, 1989), these early singing efforts are associated with involvement in social interactions, including aggression. Songs mature when the zebra finch is about 80 days old, and might be placed into use for courtship shortly after, since males mature sexually at about the same age and this species has a breeding season that extends over about seven months. An estimated 65 percent of males learn songs from their father.

What is the nature of the developmental process leading to the ability to sing normal zebra finch song (see Slater *et al.*, 1988; and Clayton, 1994)? It is restricted in time; males typically 'memorize' the songs that they will subsequently copy between 35 and 65 days of age. If separated from a singing adult male at 35 days of age and exposed to another male, a developing male will not incorporate elements from the first male's song into his own. But, if allowed to remain with an adult male until 65 days old, young males will make extensive use of the adult's song elements. Even though youngsters have fledged by the beginning of this learning phase, their continued association with their father is apparently sufficient to allow a fair amount of song learning from him (Zann, 1990). This sensitive phase for song learning does not begin and end abruptly; it varies, depending on the developing individual's circumstances. For example, males do remember the songs that they heard prior to 35 days of age, but typically do not incorporate elements from them into their own song. However, elements from early songs may be used if the social stimulation available after 35 days of age is sufficient to stimulate song development, but insufficient to stimulate song learning, e.g., because visual access to the 'tutor' is not available, or the subsequent male is a different species from the father.

The sensitive phase for song learning by zebra finches involves two separate learning processes – song memorization and learning to produce that song. However, zebra finches were not conducive to the discovery of these separate processes for two reasons: they do not readily learn song from noninteractive playbacks of recorded song, and their two stages of song development overlap in time. White-crowned sparrows differ from zebra finches with regard to both of these traits: they readily learn from playbacks of recorded song and temporally separate the two stages more than zebra finches do (e.g., see Nelson, Marler & Palleroni, 1995). As noted in Chapter 1, these sparrows have contributed significantly to the understanding of song development.

White-crowned sparrows sing one, stereotyped song. Their songs are, however, variable in another way; white-crowned sparrows exhibit dialects, in which local populations share a particular variant of the song, which is structurally distinguishable from the songs sung in adjacent populations. Figure 4.13 provides spectrograms of several white-crowned sparrow songs, and illustrates some of the differences among three identified dialects. Even though the dialect differences are evident among song pairs, the similarities among the pairs are equally conspicuous. All begin with whistles, progress to a buzz, and continue to a warbling series of note syllables or clusters. (Many also finish with another buzz, but this is soft enough to be lost in many field recordings; a terminal buzz is faintly visible in the spectrogram in Fig. 4.13c.) The variation among dialects provides a useful means to assess the extent to which vocal copy-

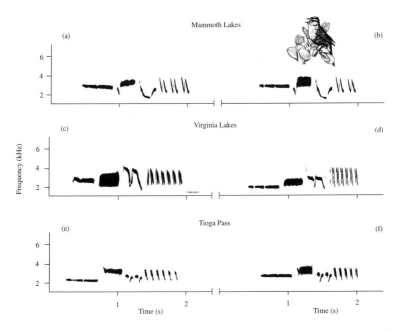

Fig. 4.13. Dialects of white-crowned sparrows from three different regions of the high Sierra Nevada mountains of California. See text for additional details. (Spectrograms and drawing courtesy of Luis Baptista.)

ing occurs; different individuals can be exposed to different dialects, and their subsequent songs can be compared to those of different dialects.

According to the early work on song development in white-crowned sparrows (Marler, 1970), memorization occurs only if males are exposed to recordings of songs during a critical phase, from 10 to 50 days of age. Delay of exposure to song playbacks until after that phase may have a general 'normalizing' effect on song structure, but the details of the training song are not copied by the male. In addition to being restricted in time, song memorization appears to involve an 'own-species' bias. White-crowned sparrow songs heard at this time become models for song production later, but the songs of other species do not. Males begin to produce song-like sounds as early as their second month post-hatching, but these subsong vocalizations are a far cry from the song of a normal adult male; no adult syllables are recognizable (Fig. 4.14a). It is not until the male makes a transition to plastic song at about 150 days of age that the effects of the memorized song begin to be evident. Plastic song is much more variable than mature song, but is divisible into introductory whistles and terminal trills, as mature song is, and contains recognizable elements of mature song (Fig. 4.14b and c). So, the memories of songs heard during the critical phase are stored without visible effect on song structure from about 50–150 days of age. This second phase appears to involve the singer comparing its actual vocal output to the structure of the memorized song, and adjusting its vocalizations to match this preferred song structure. If males are deafened after memorizing but before this 'practise' stage, then highly abnormal song is produced (Konishi, 1965). Song structure stabilizes at about 10–12 months of age (Fig. 4.14d), and little further change in vocal structure occurs.

Subsequent work indicated another, more social level of specification in the developmental systems involved in the ontogeny of song. The above work deals with specifications about song structure, indicating that the stored representations of song structure can serve as the *preferred value* in a regulatory system in which the *actual value* consists of the sounds that the bird is actually producing. Development in this case involves matching actual value to preferred value, that is, the song sung to the song memorized. But, the existence of social specifications is indicated by the fact that many species, including zebra finches, will not use noninteractive song playbacks as models in song development (Slater *et al.*, 1988); they need a singing bird to interact with. Furthermore, both the own-species bias in song structure specifications and the limitations on timing can be at least partially overridden by social circumstances.

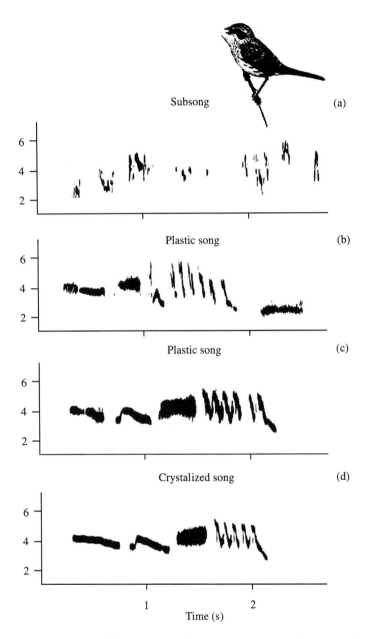

Fig. 4.14. The stages of development of song structure in white-crowned sparrows. See text for additional details. (Spectrograms courtesy of Doug Nelson.)

For example, even though zebra finch males do not typically sing the songs that they have heard prior to 35 days of age, they do remember those songs and will sing them under unusual circumstances (Clayton, 1994). If the only source of song input after 35 days is of a different species from the male who reared him, the developing male is disposed to incorporate elements of the earlier song into his own, even if the 'father' was a Bengalese finch. Similarly, if the only source of song input after 35 days is out of view, the developing male is disposed to incorporate elements of the earlier song into his own. What is deficient about input when the singer is not in view? One missing ingredient is the opportunity for social interaction. Indeed, when a young male has access to more than one singing adult, he is more likely to copy the song of the male that is more aggressive toward him. Even a tape recording of song can become a model for song learning if its playback is interactive. Male zebra finches who can turn on song playback by pecking a key copy more of the song than other males who can simply hear the playbacks but have no control over them (Adret, 1993).

Song development is typically embedded in the give and take of a social context. If we think of song and singing behavior as acts specialized to manage the behavior of others, then it is to be expected that social feedback should influence the structure of songs or the patterning of singing, or both. Developmental systems should be selected to induce ways of behaving socially that are effective in meeting the developing individual's needs.

The results of later research on song development in white-crowned sparrows were consistent with this pragmatic approach. Observations of singing behavior in natural populations revealed that some juveniles begin staking out territories in September and October, only three to four months after hatching (DeWolfe *et al.*, 1989). These intense social interactions are associated with accelerated song development, so that song crystalizes much earlier than in hand-reared birds (as early as 90 days, compared to 250+ days). Intriguingly, even prior to song crystalization, males use plastic-song in aggressive interactions, and these evoke aggressive responses by adults. Even more intriguingly, variation in plastic-song structure takes the form of singing up to four different song types. Juvenile males typically winnow such repertoires down to the one crystalized song that matches that of the males with whom they interact most. And, finally, there is evidence that plastic song is not just a developmental stepping stone to crystalized song. Young males singing plastic song will at times switch to crystalized song if challenged by another male. So, plastic song may serve a social function that is different from that of crystalized song.

Even within the species white-crowned sparrows, the details of development differ among populations in ways that appear adaptive to local conditions (Nelson *et al.*, 1995). The results described immediately above were from a nonmigratory population of the subspecies *nuttalli* living in the moderate climate of central coastal California. Other populations breed in more strongly seasonal climates, such as the subspecies *oriantha* in the high Sierra Nevada mountains of California, and they are migratory, spending only the late spring and summer in the Sierra Nevada. In laboratory studies using song playback for tutoring, *nuttalli* memorize song later on average than *oriantha* do, a possible adaptation that allows matching of song to local singers as territorial behavior begins as early as the first autumn. In contrast, *oriantha* busy themselves with migrating in the autumn, feed in flocks during the winter, and only become territorial in the next breeding season back in the mountains. Correspondingly, they produce more varieties of plastic song than *nuttalli* do, and continue in plastic song for twice as long as *nuttalli*. These *oriantha* specializations may allow yearling males better to match local songs via selective attrition of song types long after memorization is done. Such an ability may be particularly important for the males of a subspecies who move farther from their region of birth, and therefore face more uncertainty about the local song types that they will encounter.

5

Assessment/management: a viable replacement for the information concept

5.1 Why the informational perspective is inadequate

5.1.1 *Convey versus accomplish*

Major problems in the use of ideas and language have occurred in attempts to understand communication. Words often carry additional conceptual baggage that is counterproductive rather than useful. One of the difficulties about using such words is that they are liable to beg the question of causal links between process and product (Bateson, 1990). Concepts should focus attention on the processes relevant to communication. It has been seen that the concept of *information* is not equal to that task (Chapters 1–4). One reason for this is that it is not easy to ask what is *conveyed* over a longer time period than that immediately following the acoustic signal. It is, however, possible to ask what is *accomplished* over much longer periods of time. Indeed, time, long or short, must pass before accomplishment can be described.

The concept of information tends to draw attention to each signal-response couplet, thereby disposing us to treat communication as a chain of such couplets. In contrast, the regulatory *process* in A/M emphasizes the many time frames in which communicative behavior and its effects are patterned. The ultimate products of communicative behavior consist of evolutionary changes (Chapters 1–3), whereas the most proximate effects consist of the immediate impact of communication on participating individuals (Chapters 2 and 4). For both time scales, effects occur due to the interaction between processes of assessment and management (A/M). A/M describes interactions among communicants in the context of the logic of natural (and sexual) selection. Because it is a concept, A/M is not simply a research paradigm but a way of viewing the subject and of organizing research priorities. Similarly, the prevailing

228

concept for communication research, the informational perspective, is also more than simply a research paradigm.

The view that information exchange produces accomplishments is made necessary by certain assumptions that lie at the roots of informational approaches to communication. In the informational perspective, one of the major roles that conveying information plays is accounting for the *fact* of interaction between individuals. In contrast, our assumption that managing and assessing are fundamental defining properties of animals makes interaction itself a given; we do not need to account for its occurrence, but we do need to account for its form, in terms of the functions, goals, tactics, and other aspects of communication's input to the animal's fitness.

A/M starts from a premise that process and variation are fundamental properties of the world. This is consistent with a Darwinian approach, which places the process of evolutionary change and the phenomenon of interindividual variation at center stage in biology. The primacy of process in A/M also dovetails with what the philosopher Stephen Pepper has called an 'organismic' world view (Pepper, 1942). In contrast, the dependence in the informational perspective on the causal effects of information places it squarely in what Pepper calls a 'mechanistic' world view, which assumes that stability rather then change is fundamental to the world. Information exchange is needed in the informational perspective because of the assumption (often deeply buried) that the behavior of targets would not change unless some input caused it to do so. The assumption that the world is fundamentally stable makes the informational perspective less compatible with an evolutionary approach.

In an informational approach, communication becomes a series of discrete events, each consisting of a signal by one individual and a response by another. In A/M, communication reflects ongoing regulatory processes, in which patterns of different times frames, motivational concern, and outcomes are embedded. Many outcomes are balanced by conflicting sources of selection. As discussed in Chapter 2, for example, in tonic communication the short time frame of individual signal emissions functions in the longer time frame of a series of signals. Both the structure of signals and the rate of signaling influence the effect on target individuals.

Development of the concept of information was part of a cognitive revolution in the behavioral and biological sciences in which workers in those disciplines began to think of organisms as information-processing systems (Dyer, 1994). As Oyama (1985) pointed out, information was suddenly everywhere:

> In an increasingly technological, computerized world, information is a prime commodity, and when it is used in biological theorizing it is granted a kind of atomistic autonomy as it moves from place to place, is gathered, stored, imprinted and translated . . . Information, the modern source of form, is seen to reside in molecules, cells, tissues, 'the environment,' often latent but causally potent, allowing these entities to recognize, select and instruct each other, . . . to regulate, control, induce, direct, and determine events of all kinds. When something marvelous happens . . . the question is always 'Where did the information come from?'
>
> *(pp. 1–2).*

As a part of this revolution, information became a key concept in the animal behavior literature, despite some variation in its definition. All proponents of an informational perspective share the view that the impact of signals on the behavior of others is mediated by an exchange of information. This is not an effective starting place to study the process of communication because it substitutes a metaphor for a process, and consequently deflects our attention from the biological foundations responsible for the process.

For example, game theory is a popular tool to study the evolutionary dynamics of ways of participating in communication. This approach could be quite consistent with A/M; in fact, the game-theoretic idea of an evolutionary stable strategy (ESS: Maynard Smith, 1982) is essentially the same as the idea of a dynamic equilibrium in a regulatory process. (A way of behaving is evolutionarily stable if, when most members adopt it, no alternative can replace it, given current population conditions.) Unfortunately, most work with game-theoretic models has attempted to force them into the 'information transfer' paradigm. One prediction was that revealing information about one's aggressive intentions is not evolutionarily stable, but that signaling information about one's 'resource-holding potential' (roughly equal to one's 'fighting ability') is stable. As noted in Chapter 4, this prediction has its roots in a theme consistent with A/M, that assessment processes should be major sources of selection on management. Game theory predicts that (1) managers can, without constraint, exaggerate their readiness to fight, and, therefore, (2) that assessors should ignore all signals of intent because they cannot distinguish true from false signals of covert intentions. The powerful draw of informational thinking so evident in the preceding reasoning distracted researchers from the important implications of this theme, that assessment is important.

This distraction is illustrated in predictions from game theory on graded aggressive signals. Many animal signals seem graded in intensity from low to high threat, and are used sequentially in escalating agonistic encounters (Dawkins & Krebs, 1978; Bond, 1989b). A game-theoretic perspective considered these graded displays a paradox for two reasons. First, consistent with traditional emphases on the sender side of communication, they were distracted from the importance of assessment; and, second, they were paradoxically still engaging in informational thinking, by assuming that graded displays were to convey information regarding the animal's intentions. A/M predicts, instead, that if an individual escalated immediately to the highest level of threat, forgoing probing, it may find out too late that its rival is more dangerous than presumed (Markl, 1985). So, gradually escalating contests represents prudent assessment, not graded signals of the contestants' intentions!

To repeat, from the perspective of management, communication is based upon what it accomplishes, not what it conveys. The widespread usage of the informational perspective by game theorists and ethologists resulted in confusion. For example, Bond (1989a) reviews aggressive display concepts by ethologists and game theorists and concludes: 'The only reliable information about an opponent's invisible aggressive attributes, therefore, is obtained by challenging him to combat, and this essentially obviates the presumed adaptive significance of display. The result is a paradox: if displays are not truthful, they cannot have been selected for; if they were selected for communication, they cannot be truthful.'

The idea of regulation is intended to account for the surprising fact of stability in a world in flux. It is these accomplishments of regulation, the imposition of stability, that are a major concern of A/M. An analogy to some other regulatory systems shows the greater heuristic value of the A/M concept over the informational concept. Like communication, base-pair sequences in genes have long-term and short-term consequences. A molecular biologist is able to describe the base-pair sequence of a gene of interest, but unless he studies both the products of the gene and the contexts in which it is active, we are not able to say why this gene exists for we cannot assess its functional significance. For full understanding of the existence of a base pair of DNA *or* a vocal signal, it is important to know why they are not 'turned on' as well as when they are.

As mentioned, the informational perspective emphasizes the conveyance of an abstract entity in the proximate time frame. But the consequences of communication are not all proximate, as proven by the innate basis of vertebrate vocalizations (Morton, 1988). An innate, heritable, basis means that they are products of *past* accomplishments not just immediate consequences of informational couplings. Signals may or may not continue to exist, but they certainly could not exist at present without being favored by selection. Heritability leads one, logically, to study vocalizations from a form/function standpoint because they are not *de novo* creations of each individual but are products of a long history. Their form, contextual usage, and functions are subject to ongoing sources of natural selection and are not arbitrary.

Admittedly, it has been difficult to tie proximate and ultimate views of communication into a coherent whole. This is due to the obvious fact that communicating occurs nearly constantly in an individual's life, making it difficult or impossible to pinpoint the effects of communicating on reproductive success. Nonetheless, it is precisely the many causal steps between signaling and reproductive success that research strives to define.

Such attempts to tie proximate and ultimate time scales have been successful in other fields. For example, eating and looking for food are similar to communication in that they occur constantly yet are ultimately tied to reproductive payoff. Ecologists developed optimal foraging theory to circumvent the problem of 'can't see the forest for the trees' (Stephens & Krebs, 1986), and the time is ripe for those studying animal communication to attempt similar generalized descriptions more often. The discussion of song ranging (Chapter 3) and the application of signal-detection theory (Klump, 1996) are examples of such attempts. A/M is likely to foster general predictions because it is based upon evolutionary theory, rather than analogies to human communication, and it highlights regulatory processes for the proximate time frame.

To summarize, assessment/management is preferred over information for the following reasons. (1) It is not confined to particular time scales, A/M is a way of thinking that applies to all time scales, unlike the more typical treatment of proximate and ultimate as involving logically distinct processes. (2) It attributes the evolution of communication to an interplay between two processes, management and assessment, and highlights the resultant *accomplishments* for all participants, not upon what is *conveyed* immediately. Assessment and management are not synonymous with 'receiver' and 'sender' because, rather than implying that communication is a series of dyadic frames, A/M is about the interplay and the reciprocity between these roles and the regulatory processes they under-lie. Natural selection, the most powerful intellectual tool for understanding the evolution and function of communication, and the all-important answers to questions about the origins of communication systems, cannot be pushed into the back seat when A/M is the concept of choice. Even if proximate mechanisms are under study, and natural selection and evolutionary concerns are not of paramount concern, the logic of natural selection will not be violated when A/M is the concept (see T. H. Morgan's problem with evolutionary thinking in Chapter 2). (3) Unlike information, A/M fosters form and function research, thus tying ancestral to present-day communication. A/M joins communication research with other regulatory mechanisms such as foraging and competition to broaden the scientific appeal of the field. Communication joins the mainstream of modern biological and psychological science.

5.1.2 A focus on the informational perspective

The informational perspective was more radical than innovative, when it became common in the 1950s (see Fig. 1.1). No other field of biology had adopted for long a concept that would, more or less automatically, eschew a classic methodology of comparative study – relating form to function. Form and function were shown to be an integral part of communication (e.g., Marler, 1955; Morton, 1975; 1977). It remained popular to assume that the form or structure of a vocal signal was arbitrary in relation to the information, or message, it was thought to contain. The form of vocalizations was used as a label, very much like a word, but changes in form with motivation or the form itself were not appreciated (Morton, 1982). Perhaps the reason information and signal form arbitrariness has been accepted so uncritically by humans is because our speech makes use of abstract sounds (Morton & Page, 1992). Human speech may have its genetic basis but not at the level of word form (Pinker, 1994). There are notable exceptions to the arbitrariness of human speech sounds and many of these exceptions follow M–S rules, as outlined in Chapter 3 (Hinton, Nichols & Ohala, 1994).

Information is usually defined as analogous to knowledge or as a quasiquantitative 'reduction of uncertainty.' 'When we say that signs and signals convey information, we use "information" in a nonquantitative sense, although nothing in our definition would prohibit a quantitative usage. Thus, animals use signals to convey information about their identity . . ., their condition . . ., their behavioral probabilities (willing-

unable to tell males from females visually, and the males apparently did not 'recognize' their mate as an individual, at least apart from her chatter call. Furthermore, males rely upon a single call, the female chatter, to 'decide' whether or not to quit attacking. In nature, it is the female of the pair that keeps other females out of the territory and males repel other males. They, like nearly all other songbirds with permanent pair bonds, do not cooperate to defend their territory. Rather, each acts independently in its selfish interest. No wonder the males killed their mates in the newly captured pairs – they did not recognize them as females. The males used a rule of thumb – if they don't use female chatter, attack.

Why would this simple rule of thumb evolve? Some natural history is necessary at this point. A single male and a single female Carolina wren live on a territory together yearlong. They repulse same-sexed individuals from this territory. Only males sing, using about 30 versions of a loud and ringing *tea-kettle tea-kettle tea-kettle tea* throughout the day. They do not sing to attract a mate in the spring, because they already have one (Morton, 1996b). Females do not have a long-distance song, but use female chatter to indicate their sex or to threaten another female. Males and females fight with adjacent pairs but direct their attention to same-sexed birds. 'Mistakes' in sex identification reduce the effectiveness of the pair at repeling neighbors because they misdirect their attacks towards one another, wasting time and energy. This can be costly to both male and female because, unlike many bird species, wrens continually try to increase the size of their territories. Whereas males defend the territory with song, females use *chirt* calls to watch hawks, permitting the male to forage even when a hawk is perched in the territory. Both territorial expansion and female hawk surveillance serve to insure winter survival in North American populations (Morton & Shalter, 1977). Wrens are forced to forage under windfalls, which hold snow from the ground leaf litter, during heavy snowfalls. The larger the territory, the more windfalls on it. Even so, about 90 percent of wrens die during unusually snowy and cold winters. Sex identification at all times of the year is, therefore, important to reproductive success in Carolina wrens.

Most of this description of natural history and functional interpretation is not essential for an informational analysis of the communication of sex identification. Figure 5.1 depicts a diagram of wren communication of sex identification, incorporating all of the vocalizations we documented in laboratory introductions and field observations. The information provided by these displays are described first using the methods of W. J Smith (1977), and then the A/M concept is used.

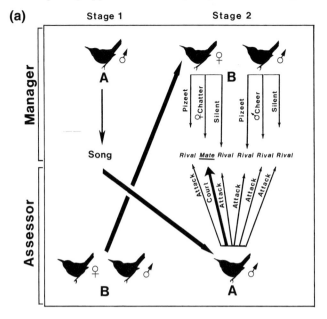

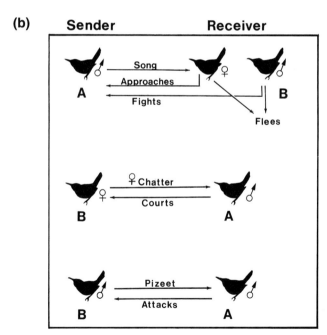

Fig. 5.1. Sex recognition through communication in the Carolina wren, diagrammed according to assessment/management (a) or informational approaches (b). See text for description.

5.2.2 *An information approach*

In Figure 5.1b, the sender (A) provides a signal containing information to make its behavior more predictable to the receiver (B) and then the receiver reciprocates. Receivers can be either male or female. In (b) the song contains the message that the sender will attack (one of the *widespread behavioral selection messages*), and a *supplemental message* that intensity will be high, as well as nonbehavioral *identifying messages* that the bird is a male and a Carolina wren. The meaning to the receiver depends on its sex, context, and its motivation. Meaning will not be dwelt on here. Although the sender's song has an attack message, the male receiver does not respond vocally until physically attacked, perhaps because the attack is accompanied by another vocalization, the *growl*, which has only an attack message with a supplemental intensity message. After the attack, a male intruder will likely flee until he is off the defender's territory. If he does not immediately try to escape (we have switched senders now, see Fig. 5.1b, lower), he may produce call notes, *pee , pi-zeet*, or male *cheer*, which carry different or shared messages. The first contains messages of escape behavior and associating behavior; it does not contain identifying messages. *Pees* occasionally grade into *scees*, with less escape message and more attack message or perhaps it conveys a general set of behavior patterns, usually alternatives physically incompatible with other behavioral selections encoded by a display (see Fig. 3.7). *Pi-zeet* contains an associating message, seeking behavior, and receptive behavior. Male *cheer* contains an attentive behavior message, indecisive behavior message, and the nonbehavioral category of identifying by specifying sex as male. In Figure 5.1b, middle, with a female present, *pees* and *pi-zeets* may be given and contain the same messages just described for a male. An attack is needed before the female vocalizes, just as is the case with a male intruder. She may produce the female chatter before the attack or even during the song. If so, she has provided a message of attack behavior and an identifying message that she is female. In Figure 5.1b, the receiver continues to attack unless the bird had uttered female chatter. In response to this message alone does the male court (see Chapter 3). Otherwise, communication usually ends at this point, as the intruder is attacked unmercifully and silently.

In another version of the informational approach, information is deduced by focusing on the receiver's responses, and functions are described based upon these responses and the contexts of the communication event. This version is more like the 'reduction in uncertainty' definition from information theory (Marler, 1961). Changes in the receiver's behavior after receipt of the signal show us that information has been received and the nature of that information. In Figure 5.1b, when the territory owner sings, the receiver might flee. Thus we interpret the information sent as aggressive. It might be more difficult to interpret the information in Figure 5.1b, middle and lower, but nonaggressive affective state is certainly salient. For both informational approaches, if the recipient had been female and she chattered, would we have interpreted the information the same? Perhaps. But we would probably be prone to describing the male's song as less aggressive, such as 'actively seizing interactional initiative' or a more aggressively neutral 'attempting to interact.' Perhaps this is more basic information to the song than aggressive information alone.

5.2.3 The assessment/management approach

The 'sender' becomes the manager and the 'receiver' becomes the assessor because these words define their communication roles (Fig. 5.1a). The regulatory problem is one common to many territorial species, to maintain oneself as the sole individual of one sex on a territory. Following initial observations, the researcher asks what tactics A might use to change B's behavior. Knowledge of the natural history suggests that male wren A, holding the territory resource, would respond to male B by repeling him such that A has exclusive use of the territory to enhance his individual survival and access to female gametes. To female B, he should improve his chances to mate by courting and behaving nonaggressively. Now we switch to present time and concentrate on the signals actually used.

Stage 1, male A produces a short song composed of three song types run together, *teakettle-cherweetie-torpedo*, then rapidly moves toward B. This song accomplishes two things: it establishes that A is a male (only male wrens sing) and the complex song makes it more likely that B, if a male, can range the close proximity of A (wrens are hard to see in their dense habitat). Singing augments the managerial impact of approaching by facilitating ranging. Stage 1 continues with production of a *growl* coincident with the attack when A is nearly upon B. This low-frequency, harsh, vocalization is a universal indicator of large size that, due to this symbolism, intimidates (remember that *growls* evolved under selection arising from the assessment rule that larger animals produce lower-frequency, broader-band sounds; see M–S rules discussion, Figs. 3.1

and 3.6, Chapter 3). Stage 2, male B produces high-frequency *peee–peee–peee* etc. and makes attempts to escape from A's claws and pecking beak. Infrequently, male B produces harsher, lower *scree* calls and simultaneously tries to fight back. The form of these sounds are diagrammed, respectively, in the upper left and right blocks in Figure 3.1. High pitch is an attempt to appease and reduce the severity of the attack by symbolizing small size, whereas the harsh addition in the *screes* reflects both aggressive and fearful endpoints combined into the same vocalization. It also might function to reduce the severity of the attack or to produce hesitation – a cornered rat will fight. Similarly, *pi-zeets* usually function in within-pair appeasement or within-brood to attract adults to feed, a high-pitched appeasement with a high-bark prefix that develops from nestling begging calls. The male *cheer* call is given by B when he is not grappling with A but results in further attacks and *growls* from A. Male *cheer* occurs in mobbing and in countersinging bouts between males. It is formed from a series of chevrons (barks) signifying that something important and interesting has been perceived but with no tendency towards either endpoint of M–S rules. These management tactics do not change the behavior of A, who continues to attack. Escape is the only option for B due to A's assessment 'rule of thumb.' None of the general appeasement vocalizations is effective in managing A and the proximate painful experience makes escape the best option for B.

Stage 2, female B is now the manager and male A is the assessor. Her use of asexual appeasement or fear-indicating calls does not change in light of A's assessment either. But this female, before A attacks and *growls*, produces a loud female chatter, characteristically a loud series of low-frequency chevrons with a harsh quality, preceded by a low bark that sounds like *nyerk*. Male A immediately metamorphoses, fluffing his feathers, spreading his tail and wings, and strutting like a miniature turkey (Fig. 5.2). If Stage 3 were illustrated in Fig. 5.1a, it would be seen that male A instantly becomes the manager again and utters high-pitched *tsucks* repeatedly and also general appeasement *pi-zeets*. He may also sing, with female B again producing female chatters, but this time she overlaps his song forming a duet.

In an A/M approach, describing the form of vocalizations is not just a convenient way of labeling them. The *pi-zeet, pee, scree, growl,* female chatter, and male *cheer* have physical form that can be placed within a model, M–S rules, to *predict* their function in management/assessment roles (see Fig. 3.1). The signal forms or physical structure become a starting point for research by generating hypotheses that can be tested

Fig. 5.2. 'Turkey' posture of a male Carolina wren.

through observation and/or experiments. There is no need to describe information as an abstract conveyance, or worse, to ascribe a causative role to information.

The hooded warbler

The male hooded warbler defended his winter territory through the use of *chinks*, a special form of bark. 'Bark' is the general term for chevron-shaped signals intermediate to the endpoints of fear and aggression discussed under M–S rules in Chapter 3. *Chinks* gave way to growls if the male needed to attack a persistent intruder (Fig. 5.3). Why didn't the male sing to defend his winter territory as he does to defend a breeding territory? A physiological explanation (hypothetical) might be that males lack high levels of testosterone in the nonbreeding season and, since song is produced under high levels, it would not occur. This may be true but we hasten to point out that our question was *why* don't they sing, not how do they! To answer this question, one needs to consider what regulatory problem the male is facing and how assessment has shaped his management strategies.

The regulatory problem The male warbler faces competition for territorial space that functions solely to provide him with food. He does not breed during the winter, so sharing the territory with a female would only

Fig. 5.3. In addition to growling when about to attack, hooded warblers show off their black throat and droop their wings, thus appearing larger than normal.

increase his chances of starving. Therefore, the regulatory problem is how to repel all other hooded warblers from his feeding territory using a minimum of energy and time to do so. Song is ruled out because it does not repel females. Barks are favored because all hooded warblers have them (both sexes) and because they are species distinctive (Morton *et al.*, 1986). Only hooded warblers will be affected by *chinks* and all can range them – they are an effective long-distance signal. By broadcasting over a large area, *chinks* save the warbler energy and time by reducing his need to patrol his large territory. The time saved can be spent looking for food. Vocalizing is probably less energy demanding than flying, and sound penetrates the dense habitat much better than optical signals. The combination of bark and species distinctiveness is the best means to announce his presence over the large territory. The growl functions to enhance the effect of aggression or threat because it symbolizes large size, as discussed in Chapter 3. The last resort is physically to eject the intruder, but such direct action is not considered communication even though it is part of the same regulatory package of territorial maintenance.

The influence of assessment   When you read the account of the hooded warbler in the Prologue you may have considered the sender in charge, 'telling' others what it wanted them to do: 'keep out!' or 'no trespassing, this property patrolled.' If you did this, your approach is a product of the informational perspective and can be contrasted with the A/M approach here. A/M focuses equally on assessing and managing, seeking insights into the success of managing by exploring the properties of assessment systems that are exploited. *Chinks* viewed separately from assessment are almost 'meaningless.' During the breeding season, they may attract, not repel, the mate because barks are often used in mobbing predators and in attracting other conspecifics (Marler, 1955). Assessors shaped the winter use of *chinks* as follows: all sexes and age classes of hooded warblers use them so they can also range them. The manager can use their ranging to threaten all hooded warblers so as to reserve the territory's resources for itself alone. It also means that assessors can use the absence of *chinks* to identify a vacant territory so owners are forced to *chink* often. Intruder pressure is high and undefended territories are invaded within minutes (Morton *et al.*, 1986). Assessment has caused managers to use *chinks* and to utter them at a high rate. As can be seen, the conclusions are different now under A/M.

We now turn to more complicated effects of assessment that occur during the male's breeding season. In the Prologue, the male hooded warbler sang one song type, a loud, clear, ringing, tonal, *weeta weeta weeTEEoo*, until he had attracted a female to his territory. At this point, the male stopped repeating the same song and mixed five other songs into his repertoire, often matching neighboring males' song types during countersinging matches. He changed from repeating a single song type, unique to him, to a mixed repertoire shared with neighboring males.

What regulatory problems are involved and how has assessment shaped management technique? The management problem is to repel males from, but attract females to, the male's territory. Many males are singing and trying to attract females so how do these females assess and choose amongst them? Females may judge both a male and his territory's quality at the same time by assessing the cumulative time he sings. If singing interferes with foraging for food, then the amount of time he spends singing provides accurate assessment (Morton, 1986). This pressure from assessment results in the male singing a great amount. It also favors his use of one song type unique to himself so that the assessor ascribes the singing correctly. After a female is acquired, she repels other females from their shared territory. Therefore, the male is unable to attract more than one female so he switches from trying to attract another to repeling neighboring males from attempting extra-pair copulations (EPCs) with his mate. Now, song output is less important than matched countersinging – singing a song type in a challenger's repertoire to threaten him through ranging effects. Still, he managed to fertilize only part of the brood he cared for because the female advertized her fertility to neighbor males by *chinking*, causing males to invade the territory to attempt EPCs (Stutchbury, Rhymer & Morton, 1994; Neudorf, Stutchbury & Piper, 1997). The regulatory problem and the vocal behavior changed from attracting to defending a female. Simultaneously, management tactics switched to exploit the assessment systems of males rather than females.

A final regulatory problem facing the hooded warbler was predator deterrence, a problem like that described for stonechats in Chapter 2. *Chinks* are, again, the vocalization best suited when a chipmunk approaches the nest because barks attract others to mob. But now the nestlings are the assessors. When the danger grew as the chipmunk climbed towards the nest, the female increased both the rate and the pitch of her *chinks*. The increased fear component produced an innate response in the nestlings, to fledge prematurely – to bail out of the dangerous nest. Assessment favored the use of *chinks* in several important events in the life history of the hooded warbler, in each case increasing the likelihood for successful management of a different regulatory problem.

Anna and her mother

Anna, the energetic 11-month-old infant, provides us with a personal glimpse of M–S rules and of the role of assessment in shaping their use in Anna's mother. Anna responds to her mother's high-pitched voice, which is expressive of small size, with signals of pleasure that her mother covets. She assesses the negative training embodied in the low 'NO!' as aversive, something to be avoided, which is why her mother, without thinking about it, used these intonations. This motherese is apparently found in all human languages (Fernald, 1992). As discussed in Chapter 2, the conative systems of infants have shaped the voice, whose inflections reflect an adaptive adult strategy to capitalize on infant conative systems. It is the infant assessor, combined with her own regulatory goals, that influences how she speaks through emotional and motivational effects on the infant. Short, rapidly repeated, broad-band notes are used to stimulate motor activity, and longer, continuous, narrow-band notes to inhibit movement. To praise an infant, mothers raise the overall pitch of their voice and vary voice pitch widely, with a rise–fall pattern. To warn their infants or express disapproval, mothers use low-pitched, short, and sharply onsetting sounds. These induce different effects in the infant by exploiting different assessment rules of thumb.

The Túngara frog

This Prologue example illustrates several principles of communication. The informational perspective seems less important to researchers studying frogs. They rarely describe the 'information' transmitted by frog mating calls but, instead, describe function, perception, and evolutionary aspects of their subject. One result of this approach is the description of sources of selection affecting communication in frogs that is unparalleled. One principle the Prologue example should bring to mind is the importance of eavesdropping rather than social consequences to communication – witness the fringe-lipped bat cueing in on the *chucks* of male courtship songs! Predation from this bat has selected for the ability of males to leave *chucks* off the whine call. Another principle is that of size/sound relationships. In this case, females are more stimulated by low than by higher-pitched *chucks*. Because lower pitch is related to larger size, females thereby show a preference to mate with the largest males, those having the lowest-pitched *chucks*.

This is a good example of how A/M transcends the proximate/ultimate dichotomy. Female frogs are engaged in the proximate activity of choosing mates based upon the evolutionary principle that large males have better genes because they must have survived longer to get larger. To the extent that survival time is heritable, females pass on better genes to their offspring by basing mate choice on male size. Female assessment is responsible for the use of chucks by males, even though bats may use them with opposite fitness results for the males (Ryan, 1985). This is a case of natural selection and sexual selection having provided opposite sources of selection on the males' *whiine, chuck-chuck; whiine, chuck-chuck* advertizement call. Obviously, assessment by females overcomes the assessment through eavesdropping by predatory bats in the Túngara frog, *Physalaemus pustulosus*.

But there is more to this story than was presented in the Prologue. There are many other species in the genus *Physalaemus* and many of them do not have the *chuck* but only the *whine* portion of the advertizement call. Ryan and Rand (1993a; 1993b; 1993c) performed a series of mate-choice experiments using a two-speaker playback technique. Two different stimuli are played from the two speakers with the female placed between them. Females respond by approaching one or the other of the speakers and this is considered her choice. Ryan and Rand determined, not unexpectedly, that females of each species of *Physalaemus* chose their own species-specific call. But when they added *chucks* of *pustulosus* to the *whine* of *coloradorum*, a species that does not have the *chuck* and does not overlap geographically with *pustulosus*, female *coloradorum* preferred the modified call over their normal *chuckless whine*. The species-distinctive *whine*, when coupled with the *chuck* from another species (something the female *coloradorum* had never heard), resulted in a preference! Ryan and Rand suggested that their data illustrated a bias in the females' sensory system, in this case not exploited by male *coloradorum* (perhaps due to a higher level of predation that shifted the balance of selection against the use of the *chuck*). Males could exploit such a physiologically based bias in their competition with one another for mates. The general notion of such exploitation they called a sensory exploitation. This idea was first introduced in Chapter 1.

Sensory exploitation suggests that females exhibit preferences for traits that do not yet exist in males. It follows that males have the opportunity to tap into these sensory biases, that is, to respond to the assessors' biases the better to affect their assessment in a way favorable to the mating success of individual males. Work with other species supports the notion of sensory exploitation, that sensory biases exist that predate the appearance of a sexually selected trait in the population. Water mites (Proctor, 1993), fiddler crabs (Christy, 1988), *Anolis* lizards (Fleischman, 1992), and grackles (Searcy, 1992) are examples. The idea of sensory exploitation illustrates how syntheses of proximate and ultimate approaches can be achieved. It is the understanding of the properties of proximate assessment mechanisms that can provide answers to ultimate questions such as why vocalizations have evolved the forms that they have.

Sensory exploitation is another example of the dynamic view of communication espoused by A/M. Its novelty is more apparent than real due to misconceptions from informational thinking about communication. When the role of assessment in channeling management is highlighted, sensory exploitation by managers is a logical prediction as a solution to their regulatory problems.

The California ground squirrel and the rattlesnake

This Prologue example was founded on research illustrating several implications of an A/M approach. Perhaps most important, it provided an example of active assessment processes. As discussed in Chapters 2 and 4, ground squirrels need to know both how warm and how big the rattlesnakes are that they encounter, because larger and warmer snakes are more dangerous. Visual cues about size are not available in the darkness of the burrow, and are not all that useful anyway for assessing snake temperature. The female's blast of sand at the snake appears to function to elicit rattling by the snake, a sound that includes cues about both snake size and temperature. This feedback is very useful to maternal ground squirrels; playback studies demonstrated that they can use the sound alone to judge the level of risk that they face.

The 'informativeness' of rattling by rattlesnakes helps us to understand in general how it is that signals come to *seem specialized to make information available*, even though that is not what signals are specialized to do at all. The availability of cues about snake size and temperature in rattling does not even reflect selection for more effective signal function, let alone selection to make information available. Variation in rattling is no richer a source of cues about differences in snake dangerousness than the nonsignaling act of striking is (see Fig. 2.12). Rattling and striking are similar in having become potential sources of assessment cues through the physical and physiological factors that constrain and account for variation in their form. The body size and body temperature cues available in rattling have their origins in relatively straightforward physical and physiological constraints. For example, larger snakes can strike with higher velocity, and the dominant frequency of their rattling sounds is lower because their rattles are larger and therefore have lower resonant frequencies. Similarly, colder snakes strike more slowly and rattle with lower click rates because muscle contraction speed declines as muscle temperature drops. From an A/M perspective, the 'informativeness' of signals arises from the constraints of signaling situations, especially from the feedback generated by the assessment systems of targets.

Finally, A/M emphasizes that the interplay between assessment and management persists not only through the interaction of existing participants in the communication system, but also as a result of intrusion by 'outside' third parties. Such outside parties can include predators, who find prey by eavesdropping on their signals (Túngara frogs and fringe-lipped bats), and mimics, who gain by exploiting the evocativeness of a signal. Rattling is evocative because many animals know of its association with venomousness; burrowing owls protect themselves in their burrows by producing vocalizations virtually identical to the rattling sound of rattlesnakes (Chapter 2).

5.3 Assessment and management under natural and sexual selection

One reason that sexual selection, a form of natural selection, is discussed as a special case of selection is that it may oppose the effects of other sources of natural selection and result in a reduction in survival (Andersson, 1994; Tanaka, 1996). As has just been seen, the *chucks* in Túngara frog calls are such a case. Do sexual and natural selection also produce consistent differences in A/M outcomes for communication?

5.3.1 Sexual selection

Sexual selection, i.e., selection for traits which are solely concerned with increasing mating success, can work either by favoring males who compete with other males for fertilizations more successfully, or by favoring traits in one sex that attract the other, or both. Male–male competition often involves overt aggression because the function is to procure a resource (e.g., a territory or oviposition site) and other males have equally selfish needs. The assessors (females) are probably one step removed from the control of the resources they need; it is still debatable, for example, whether a female songbird assesses the quality of a territory and accepts any male possessing it or assesses the male's quality apart from his territory. Perhaps both occur (e.g., Mountjoy & Lemon, 1996). But the importance of female assessment (called choosiness in sexual selection literature) selects for males to acquire resources that the females need in order to reproduce. Females may obtain other direct benefits from males, such as parental care and resource provisioning, which influence their assessment of potential mates. Aggression amongst males in male–male competition is often high because it is less likely that incontestable signals can substitute for aggressive contests. The stakes are too high. Probing brings about immediate escalation by managers.

On the other hand, female mate choice is most likely to favor communication, and often leads to its most bizarre and extravagant forms, like peacock tails and bowerbird wails (Loffredo & Borgia, 1986). And the most extravagant signals are found in those mating systems where the female receives only sperm from her mate. Here, assessment of the male takes place without any other resource to assess but his genetic quality. Incontestable signals must reflect a commodity, genetic quality, that cannot be seen or measured directly in this situation. Leks, polygamous mating systems, and extra-pair mating systems are examples where the male provides only sperm to females. Females are said to obtain indirect benefits that can arise from arbitrary male attractiveness (Fisher, 1930; Lande, 1981) or viability signaled by a secondary sexual character (Zahavi, 1975; Kodric-Brown & Brown, 1984; Andersson, 1986; Davidar & Morton, 1993). Often, under these circumstances,

References

Acredolo, L. P. & Goodwyn, S. W. (1990). Sign language in babies: The significance of symbolic gesturing for understanding language development. In *Annals of Child Development*, 7, ed. R. Vasta, pp. 1–42. London: Jessica Kingsley Publishers.

Adams, E. S. & Caldwell, R. L. (1990). Deceptive communication in asymmetric fights of the stomatopod crustacean *Gonodactylus bredini*. *Animal Behaviour*, **39**, 706–16.

Adret, P. (1993). Operant conditioning, song learning and imprinting to taped song in the zebra finch. *Animal Behaviour*, **46**, 149–59.

Agmo, A. & Berenfeld, R. (1990). Reinforcing properties of ejaculation in the male rat: Role of opioids and dopamine. *Behavioral Neuroscience*, **104**, 177–82.

Alatalo, R. V., Carlson, A., Lundberg, A. & Ulfstrand, S. (1981). The conflict between male polygamy and female monogamy: the case of the pied flycatcher, *Ficedula hypoleuca*. *American Naturalist*, **117**, 738–53.

Alatalo, R. V., Glynn, C. & Lundberg, A. (1990). Singing rate and female attraction in the pied flycatcher: an experiment. *Animal Behaviour*, **39**, 601–3.

Alcock, J. (1989). *Animal Behavior: An Evolutionary Approach*. Sunderland: Sinauer Associates.

Anderson, C. O. & Mason, W. A. (1974). Early experience and complexity of social organization in groups of young rhesus monkeys (*Macaca mulatta*). *Journal of Comparative and Physiological Psychology*, **87**, 681–90.

Andersson, M. (1986). Evolution of condition-dependent sex ornaments and mating preferences: Sexual selection based on viability differences. *Evolution*, **40**, 804–16.

Andersson, M. (1994). *Sexual Selection*. Princeton: Princeton University Press.

Andrew, R. J. (1972). The information potentially available in mammalian displays. In *Non-Verbal Communication*, ed. R. A. Hinde, pp. 179–204. Cambridge: Cambridge University Press.

Arak, A. (1983). Sexual selection by male–male competition in natterjack toad choruses. *Nature*, **306**, 261–2.

Archer, J. (1988). *The Behavioural Biology of Aggression*. Cambridge: Cambridge University Press.

Arnold, A. P., Bottjer, S. W., Brenowitz, E. A., Nordeen, E. J. & Nordeen, K. W. (1986). Sexual dimorphisms in the neural vocal control system in song birds: ontogeny and phylogeny. *Brain, Behavior and Evolution*, **28**, 22–31.

Arnold, S. J. (1994). Constraints on phenotypic evolution. In *Behavioral Mechanisms in Evolutionary Ecology*, ed. L. A. Real, pp. 258–78. Chicago: University of Chicago Press.

Arrowood, P. C. (1988). Duetting, pair bonding and agonistic display in parakeet pairs. *Behaviour*, **106**, 129–57.

Arvidsson, B. L. & Neergaard, R. (1991). Mate choice in the willow warbler – a field experiment. *Behavioral Ecology Sociobiology*, **29**, 225–9.

Aubin, T. (1987). Respective parts of the carrier and of the frequency modulation in the semantics of distress calls. *Behaviour*, **100**, 123–33.

Aubin, T. (1991). Why do distress calls evoke interspecific responses? An experimental study applied to some species of birds. *Behavioural Processes*, **23**, 103–11.

August, P. V. & Anderson, J. G. T. (1987). Mammal sounds and motivation–structural rules: a test of the hypothesis. *Journal of Mammalogy*, **68**, 1–9.

Austen, M. J. W. & Handford, P. T. (1991). Variation in the songs of breeding Gambel's white-crowned sparrows near Churchill, Manitoba. *Condor*, **93**, 147–52.

Baker, M. C. (1974). Genetic structure of two populations of white-crowned sparrows with different song dialects. *Condor*, **76**, 351–6.

Baptista, L. F. (1975). Song dialects and demes in sedentary populations of the white-crowned sparrow, (*Zonotrichia leucophrys nuttali*). *University of California Publications in Zoology*, **105**, 1–52.

Baptista, L. F. (1978). Territorial, courtship and duet songs of the Cuban grassquit (*Tiaris canora*). *Journal für Ornithologie*, **119**, 91–101.

Baptista, L. F. (1985). The functional significance of song sharing in the White-crowned sparrow. *Canadian Journal of Zoology*, **63**, 1741–52.

Baptista, L. F. & Gaunt, S. L. L. (1994). Historical perspectives: Advances in studies of avian sound communication. *Condor*, **96**, 817–30.

Baptista, L. F. & Petrinovich, L. (1984). Social interaction, sensitive phases and the song template hypothesis in the white-crowned sparrow. *Animal Behaviour*, **32**, 172–81.

Baptista, L. F. & Schuchmann, K. L. (1990). Song learning in the Anna hummingbird (*Calypte anna*). *Ethology*, **84**, 15–26.

Baptista, L. F. & Trail, P. W. (1992). The role of song in the evolution of passerine diversity. *Systematic Biology*, **41**, 242–7.

Barlow, G. W., Rogers, W. & Fraley, N. (1986). Do midas cichlids win through prowess or daring? It depends. *Behavioral Ecology and Sociobiology*, **19**, 1–8.

Bateson, P. (1987). Imprinting as a process of competitive exclusion. In *Imprinting and Cortical Plasticity: Comparative Aspects of Sensitive Periods. Wiley Series in Neuroscience*, **1**, ed. J. P. Rauschecker and P. Marler, pp. 151–68. New York: John Wiley & Sons.

Bateson, P. (1990). Choice, preference, and selection. In *Interpretation and Explanation in the Study of Animal Behavior*, **1**, ed. M. Bekoff and D. Jamieson, pp. 149–56. Boulder: Westview Press.

Beer, C. G. (1973). A view of birds. *Minnesota Symposia on Child Psychology*, **7**, 47–86.

Beer, C. G. (1975). Multiple functions and gull displays. In *Function and Evolution in Behaviour*, ed. G. Baerends, C. Beer & A. Manning, pp. 16–54. Oxford: Clarendon Press.

Beer, C. G. (1977). What is a display? *American Zoologist*, **17**, 155–65.

Beer, C. (1982). Conceptual issues in the study of communication. In *Acoustic Communication in Birds, Vol. 2, Song Learning and its Consequences*, ed. D. E. Kroodsma & E. H. Miller, pp. 279-310. New York: Academic Press.

Beletsky, L. D. (1983). Aggressive and pair-bonded maintenance songs of female red-winged blackbirds (*Agelaius phoeniceus*). *Zeitschrift für Tierpsychologie*, **62**, 47–54.

Beletsky, L. D. (1989). Communication and the cadence of birdsong. *American Midland Naturalist*, **122**, 298–306.

Bench, J. (1969). Some effects of audio-frequency stimulation on the crying baby. *Journal of Auditory Research*, **9**, 122–8.

Berger, J. (1981). The role of risks in mammalian combat: Zebra and onager fights. *Zeitschrift für Tierpsychologie*, **56**, 297–304.

Berger, J. (1986). *Wild Horses of the Great Basin: Social Competition and Population Size*. Chicago: University of Chicago Press.

Bernieri, F. J., Reznick, J. S. & Rosenthal, R. (1988). Synchrony, pseudosynchrony, and dissynchrony: Measuring the entrainment process in mother–infant interactions. *Journal of Personality and Social Psychology*, **54**, 243–53.

Berridge, K. C. (1996). Food reward – brain substrates of wanting and liking. *Neuroscience and Biobehavioral Reviews*, **20**, 1–25.

Bertram, B. (1970). The vocal behaviour of the Indian hill mynah, *Gracula religiosa*. *Animal Behaviour Monograph*, **8**, 81–192.

Bilger, R. C. & Hirsh, I. J. (1956). Masking of tones by bands of noise. *Journal of the Acoustical Society of America*, **28**, 623–30.

Birns, B., Blank, M., Bridger, W. & Escalona, S. (1965). Behavioral inhibition in neonates produced by auditory stimuli. *Child Development*, **36**, 639–45.

Bischof, H.-J. (1994). Sexual imprinting as a two-stage process. In *Causal Mechanisms of Behavioural Development*, ed. J. A. Hogan & J. J. Bolhuis, pp. 82-97. Cambridge: Cambridge University Press.

Bischof, H.-J. & Clayton, N. (1991). Stabilization of sexual preferences by sexual experience in male zebra finches *Taeniopygia guttata castanotis*. *Behaviour*, **118**, 144–55.

Bitterbaum, E. & Baptista, L. F. (1979). Geographical variation in songs of California house finches (*Carpodacus mexicanus*). *Auk*, **96**, 462–74.

Blumberg, M. S. & Alberts, J. R. (1991). On the significance of similarities between ultrasonic vocalizations of infant and adult rats. *Neuroscience and Biobehavioral Reviews*, **15**, 383–90.

Blumberg, M. S. & Alberts, J. R. (1997). Incidental emissions, fortuitous effects, and the origins of communication. In *Communication. Perspectives in Ethology*, **12**, ed. D. H. Owings, M. D. Beecher & N. S. Thompson. New York: Plenum Press.

Boellstorff, D. E., Owings, D. H., Penedo, M. C. T. & Hersek, M. J. (1994). Reproductive behaviour and multiple paternity of California ground squirrels. *Animal Behaviour*, **47**, 1057–64.

Bogert, C. M. (1960). The influence of sound on amphibians and reptiles. In *Animal Sounds and Communication*, ed. W. E. Lanyon & W. Tavolga, pp. 137–320. Washington: American Institute of Biological Science.

Bolhuis, J. J. (1991). Mechanisms of avian imprinting: a review. *Biological Reviews*, **66**, 303–45.

Bolhuis, J. J. & Van Kampen, H. S. (1992). An evaluation of auditory learning in filial imprinting. *Behaviour*, **122**, 195–230.

Bolles, R. C. (1970). Species-specific defense reactions and avoidance learning. *Psychological Review*, **77**, 32–48.

Bond, A. B. (1989a). Toward a resolution of the paradox of aggressive displays: I. Optimal deceit in the communication of fighting ability. *Ethology*, **81**, 29–46.

Bond, A. B. (1989b). Toward a resolution of the paradox of aggressive displays: II. Behavioral efference and the communication of intentions. *Ethology*, **81**, 235–49.

Bossema, I. & Burgler, R. R. (1980). Communication during monocular and binocular looking in European jays, (*Garrulus g. glandiarus*). *Behaviour*, **74**, 274–83.

Bowlby, J. (1969). *Attachment*. New York: Basic Books.

Bowman, R. I. (1979). Adaptive morphology of song dialects in Darwin's finches. *Journal für Ornithologie*, **120**, 353–89.

Bremond, J. C. (1978). Acoustic competition between the song of the wren (*Troglodytes troglodytes*) and the songs of other species. *Behaviour*, **65**, 89.

Brenowitz, E. A. (1982). Long-range communication of species identity by song in the red-winged blackbird. *Behavioral Ecology and Sociobiology*, **10**, 29–38.

Brooks, R. J. & Falls, J. B. (1975). Individual recognition by song in white-throated sparrows: I. Discrimination of songs of neighbors and strangers. *Canadian Journal of Zoology*, **53**, 879–88.

Brown, A. M. (1973). High levels of responsiveness from the inferior colliculus of rodents at ultrasonic frequencies. *Journal of Comparative Physiology*, **83**, 393–406.

Brown, C. H. (1982). Ventriloquial and locatable vocalizations in birds. *Zeitschrift für Tierpsychologie*, **59**, 338–50.

Burghardt, G. M. (1970). Defining 'communication'. In *Communication by Chemical Signals. Advances in Chemoreception*, **1**, ed. J. W. Johnston Jr, D. G. Moulton & A. Turk, pp. 5–18. New York: Appleton-Century-Crofts

Byrne, J. M. & Horowitz, F. D. (1981). Rocking as a soothing intervention: The influence of direction and type of movement. *Infant Behavior and Development*, **4**, 207–18.

Caine, N. G., Addington, R. L. & Windfelder, T. L. (1995). Factors affecting the rates of food calls given by red-bellied tamarins. *Animal Behaviour*, **50**, 53–60.

Caldwell, R. L. (1986). The deceptive use of reputation by stomatopods. In *Deception: Perspectives on Human and Nonhuman Deceit*, ed. R. W. Mitchell & N. S. Thompson, pp. 129–45. Albany: State University of New York Press.

Cannon, W. B. (1935). Stresses and strains of homeostasis. *The American Journal of the Medical Sciences*, **189**, 1–14.

Capp, M. S. & Searcy, W. A. (1991). Acoustical communication of aggressive intentions by territorial male bobolinks. *Behavioral Ecology*, **2**, 319–26.

Capranica, R. R., Frishkopf, L. S. & Nevo, E. (1973). Encoding of geographic dialects in the auditory system of the cricket frog. *Science*, **182**, 1272–5.

Capranica, R. R. & Moffat, A. J. M. (1983). Neurobehavioral correlates of sound communication in anurans. In *Advances in Vertebrate Neuroethology*, ed. J. P. Ewert, R. R. Capranica & D. J. Ingle, pp. 701–30. New York: Plenum Press.

Caro, T. M. & Hauser, M. D. (1992). Is there teaching in nonhuman animals? *Quarterly Review of Biology*, **67**, 151–74.

Catchpole, C. K. & Leisler, B. (1989). Variation in the song of the aquatic warbler *Acrocephalus paludicola* in response to playback of different song structures. *Behaviour*, **108**, 125–38.

Catchpole, C. K. & Slater, P. J. B. (1995). *Bird Song, Biological Themes and Variations*. Cambridge: Cambridge University Press.

Chappuis, C. (1971). Un exemple de l'enfluence du milieu sur les emissions vocales des oiseaux: l'evolution des chants en foret equitoriale. *Terre et Vie*, **118**, 183–202.

Cheney, D. L. & Seyfarth, R. M. (1985). Vervet monkey alarm calls: manipulation through shared information? *Behaviour*, **94**, 150–66.

Cheney, D. L. & Seyfarth, R. M. (1988). Assessment of meaning and the detection of unreliable signals by vervet monkeys. *Animal Behaviour*, **36**, 477–86.

Cheney, D. L. & Seyfarth, R. M. (1990). *How Monkeys See the World: Inside the Mind of Another Species*. Chicago: University of Chicago Press.

Cheney, D. L., Seyfarth, R. M. & Silk, J. B. (1995). The role of grunts in reconciling opponents and facilitating interactions among adult female baboons. *Animal Behaviour*, **50**, 249–57.

Cheng, M.-F. (1986). Female cooing promotes ovarian development in ring doves. *Physiology and Behavior*, **37**, 371–4.

Cheng, M. F. (1992). For whom does the female dove coo? A case for the role of vocal self-stimulation. *Animal Behaviour*, **43**, 1035–44.

Cherry, C. (1957). *On Human Communication: A Review, A Survey, and A Criticism*. Cambridge: Technology Press of MIT.

Cherry, C. (1966). *On Human Communication*. Cambridge: MIT Press.

Christy, J. G. (1988). Pillar function in the fiddler crab *Uca beebi*: 1. Competitive courtship signalling. *Ethology*, **78**, 113–28.

Clayton, N. S. (1994). The influence of social interactions on the development of song and sexual preferences in birds. In *Causal Mechanisms of Behavioural Development*, ed. J. A. Hogan & J. J. Bolhuis, pp. 98–115. Cambridge: Cambridge University Press.

Clutton-Brock, T. H. & Albon, S. D. (1979). The roaring of red deer and the evolution of honest advertisement. *Behaviour*, **69**, 145–70.

Clutton-Brock, T. H., Guinness, F. E. & Albon, S. D. (1982). *Red Deer: Behavior and Ecology of Two Sexes*. Chicago: University of Chicago Press.

Clutton-Brock, T. H. & Parker, G. A. (1995). Punishment in animal societies. *Nature*, **373**, 209–16.

Coe, C. L. (1990). Psychobiology of maternal behavior in nonhuman primates. In *Mammalian Parenting: Biochemical, Neurobiological, and Behavioral Determinants*, ed. N. A. Krasnegor & R. S. Bridges, pp. 157–83. New York: Oxford University Press.

Cohen, G. (1990). Data-driven and conceptually-driven processing. In *The Blackwell Dictionary of Cognitive Psychology*, ed. M. W. Eysenck, pp. 88–9. Oxford: Basil Blackwell.

Collias, N. E. (1963). A spectrographic analysis of the vocal repertoire of the African village weaverbird. *Condor*, **65**, 517–27.

Cosens, S. E. & Falls, J. B. (1984a). Structure and use of song in the yellow-headed blackbird (*Xanthocephalus xanthocephalus*). *Zeitschrift für Tierpsychologie*, **66**, 227–41.

Cosens, S. E. & Falls, J. B. (1984b). A comparison of sound propagation and song frequency in temperate marsh and grassland habitats. *Behavioral Ecology and Sociobiology*, **15**, 161–70.

Coss, R. G. (1991). Evolutionary persistence of memory-like processes. *Concepts in Neuroscience*, **2**, 129–68.

Coss, R. G., Guse, K. L., Poran, N. S. & Smith, D. G. (1993). Development of antisnake defenses in California ground squirrels (*Spermophilus beecheyi*): II. Microevolutionary effects of relaxed selection from rattlesnakes. *Behaviour*, **124**, 137–64.

Coss, R. G. & Owings, D. H. (1985). Restraints on ground squirrel antipredator behavior: Adjustments over multiple time scales. In *Issues in the Ecological Study of Learning*, ed. T. D. Johnston & A. T. Pietrewicz, pp. 167–200. Hillsdale, NJ: Lawrence Erlbaum Associates.

Craig, J. L. & Jenkins, P. F. (1982). The evolution of complexity in broadcast songs of passerines. *Journal of Theoretical Biology*, **95**, 415–22.

Craig, W. (1918). Appetites and aversions as constituents of instinct. *Biological Bulletin*, **34**, 91–107.

Cronin, H. (1991). *The Ant and the Peacock: Altruism and Sexual Selection from Darwin to Today*. Cambridge: Cambridge University Press.

Curio, E., Ernst, U. & Vieth, W. (1978). Cultural transmission of enemy recognition: One function of mobbing. *Science*, **202**, 899–901.

Currie, P. J. & Sarjeant, W. A. S. (1979). Lower Cretaceous dinosaur footprint from the Peace River Canyon, British Columbia, Canada. *Palaeogeography, Palaeoclimatology, Palaeoecology*, **28**, 103–15.

Cuthill, I. & Macdonald, W. A. (1990). Experimental manipulation of the dawn and dusk chorus in the blackbird (*Turdus merula*). *Behavioral Ecology and Sociobiology*, **26**, 209–16.

Dabelsteen, T. (1992). Interactive playback: a finely tuned response. In *Playback and Studies of Animal Communication*, ed. P. K. McGregor, pp. 97–109. New York: Plenum Press.

Dabelsteen, T. & Pedersen, S. B. (1990). Song and information about aggressive responses of blackbirds, *Turdus merula*: evidence from interactive playback experiments with territory owners. *Animal Behaviour*, **40**, 1158–68.

Dale, S. & Slagsvold, T. (1994). Polygyny and deception in the pied flycatcher: can females determine male mating status? *Animal Behaviour*, **48**, 1207–17.

Darwin, C. (1871/1981). *The Descent of Man, and Selection in Relation to Sex*. Princeton: Princeton University Press.

Darwin, C. (1872/1965). *The Expression of the Emotions in Man and Animals*. Chicago: University of Chicago Press.

Davidar, P. & Morton, E. S. (1993). Living with parasites: Prevalence and effects of a blood parasite on survivorship in the Purple Martin. *Auk*, **110**, 109–16.

Davies, N. B. & Halliday, T. R. (1978). Deep croaks and fighting assessment in toads, *Bufo bufo*. *Nature*, **274**, 683–5.

Davis, M. (1982). *Interaction Rhythms: Periodicity in Communicative Behavior*. New York: Human Sciences Press.

Dawkins, M. S. & Guilford, T. (1991). The corruption of honest signalling. *Animal Behaviour*, **41**, 865–73.

Dawkins, R. (1976). *The Selfish Gene*. Oxford: Oxford University Press.

Dawkins, R. & Krebs, J. R. (1978). Animal signals: Information or manipulation? In *Behavioural Ecology: An Evolutionary Approach*, ed. J. R. Krebs & N. B. Davies, pp. 282–309. Sunderland: Sinauer Associates.

Dawkins, R. & Krebs, J. R. (1979). Arms races between and within species. *Proceedings of the Royal Society of London, Series B*, **205**, 489–511.

Dawson, S. M. (1982). On the evolution of song repertoire. A discussion of the evidence. *Tuatara*, **26**, 27–35.

Decoursey, P. (1961). Effect of light on the circadian activity rhythm of the flying squirrel *Glaucomys volans*. *Zeitschrift für Vergleichende Physiologie*, **44**, 331–54.

Dennett, D. C. (1983). Intentional systems in cognitive ethology: The 'Panglossian paradigm' defended. *Behavioral and Brain Sciences*, **6**, 343–90.

DeWolfe, B. B., Baptista, L. F. & Petrinovich, L. (1989). Song development and territory establishment in Nuttall's White-crowned Sparrows. *Condor*, **91**, 397–407.

Dewsbury, D. A. (1992). On the problems studied in ethology, comparative psychology, and animal behavior. *Ethology*, **92**, 89–107.

Dickinson, A. (1980). *Contemporary Animal Learning Theory*. Cambridge: Cambridge University Press.

Dingle, H. (1969). A statistical and information analysis of aggressive communication in the mantis shrimp *Gonodactylus bredini* Manning (Crustacea: Stomatopoda). *Animal Behaviour*, **17**, 561–75.

Dixson, A. F. (1980). Androgens and aggressive behavior in primates: A review. *Aggressive Behavior*, **6**, 37–67.

Dooling, R. J. (1982). Auditory perception in birds. In *Acoustic Communication in Birds*, **1**, ed. D. E. Kroodsma & E. H. Miller, pp. 95–130. New York: Academic Press.

Driver, P. M. & Humphries, D. A. (1969). The significance of the high-intensity alarm call in captured passerines. *Ibis*, **111**, 243–4.

Duellman, W. E. & Pyles, R. A. (1983). Acoustic resource partitioning in anuran communities. *Copeia*, 639–49.

Dyer, F. C. (1994). Spatial cognition and navigation in insects. In *Behavioral Mechanisms in Evolutionary Ecology*, ed. L. A. Real, pp. 66–98. Chicago: University of Chicago Press.

Ekman, P. (1992). Facial expressions of emotion: an old controversy and new findings. *Philosophical Transactions of the Royal Society of London B*, **335**, 63–9.

Elowson, A. M., Tannenbaum, P. L. & Snowdon, C. T. (1991). Food-associated calls correlate with food preferences in cotton-top tamarins. *Animal Behaviour*, **42**, 931–7.

Emlen, S. T. (1971). The role of song in individual recogniton in the indigo bunting. *Zeitschrift für Tierpsychologie*, **28**, 241–6.

Emlen, S. T., Wrege, P. H. & Demong, N. J. (1995). Making decisions in the family: an evolutionary perspective. *American Scientist*, **83**, 148–57.

Endler, J. A. (1992). Signals, signal conditions, and the direction of evolution. *American Naturalist*, **139**, S125–53.

Enquist, M. (1985). Communication during aggressive interactions with particular reference to variation in choice of behaviour. *Animal Behaviour*, **33**, 1152–61.

Evans, C. (1997). Referential signals. In *Communication. Perspectives in Ethology*, **12**, ed. D. H. Owings, M. D. Beecher & Thompson. New York: Plenum Press.

Evans, R. M. (1990a). Embryonic fine tuning of pipped egg temperature in the American white pelican. *Animal Behaviour*, **40**, 963–8.

Evans, R. M. (1990b). Vocal regulation of temperature by avian embryos: a laboratory study with pipped eggs of the American white pelican. *Animal Behaviour*, **40**, 969–79.

Evans, R. M. (1992). Embryonic and neonatal vocal elicitation of parental brooding and feeding responses in American white pelicans. *Animal Behaviour*, **44**, 667–75.

Evans, R. M. (1994). Cold-induced calling and shivering in young American white pelicans: honest signalling of offspring need for warmth in a functionally integrated thermoregulatory system. *Behaviour*, **129**, 13–34.

Falls, J. B. (1982). Individual recognition by sound in birds. In *Acoustic Communication in Birds*, **2**, ed. D. E. Kroodsma & E. H. Miller, pp. 237–73. New York: Academic Press.

Fanselow, M. S. (1989). The adaptive function of conditioned defensive behavior: An ecological approach to Pavlovian stimulus-substitution theory. In *Ethoexperimental Approaches to the Study of Behavior. NATO Advanced Science Institutes Series. Series D: Behavioural and Social Sciences*, **48**, ed. R. J. Blanchard, P. F. Brain, D. C. Blanchard & S. Parmigiani, pp. 151–66. Dordrecht, Netherlands: Kluwer Academic Publishers.

Fanselow, M. S. (1991). Analgesia as a response to aversive Pavlovian conditional stimuli: cognitive and emotional mediators. In *Fear, Avoidance and Phobias: a Fundamental Analysis*, ed. M. R. Denny, pp. 61–86. Hillsdale, NJ: Lawrence Erlbaum Associates.

Fenton, M. B. (1994). Assessing signal variability and reliability: 'to thine self be true'. *Animal Behavior*, **47**, 757–64.

Fernald, A. (1992). Human maternal vocalizations to infants as biologically relevant signals: An evolutionary perspective. In *The Adapted Mind: Evolutionary Psychology and the Generation of Culture*, ed. J. H. Barkow, L. Cosmides & J. Tooby, pp. 345–82. Oxford: Oxford University Press.

Ficken, M. S., Ficken, R. W. & Witkin, S. R. (1978). Vocal repertoire of the Black-capped Chickadee. *Auk*, **95**, 34–48.

Ficken, R. W., Ficken, M. S. & Hailman, J. P. (1974). Temporal pattern shifts to avoid acoustic interference in singing birds. *Science*, **183**, 762–3.

Fisher, J. (1954). Evolution and bird sociality. In *Evolution as a Process*, ed. J. Huxley, C. Hardy & E. B. Ford, pp. 71–83. London: Allen & Unwin.

Fisher, R. A. (1930). *The Genetical Theory of Natural Selection*. Oxford: Clarendon Press.

Fitch, H. S. (1949). Study of snake populations in central California. *American Midland Naturalist*, **41**, 513–79.

Fleischman, L. (1992). The influence of the sensory system and the environment on motion patterns in the visual displays of anoline lizards and other vertebrates. *American Naturalist*, **139**, S36–S61.

Fletcher, H. & Munson, W. A. (1937). Relation between loudness and masking. *Journal of the Acoustical Society of America*, **9**, 1–10.

Folger, T. (1993). The blood of the dinos. *Discover*, **14**, 99.

Frazer, G., Sissom, D., Rice, D. & Peters, G. (1991). How cats purr. *Journal of the Zoological Society of London*, **223**, 67–78.

Fridlund, A. J. (1994). *Human Facial Expression: An Evolutionary View*. San Diego: Academic Press.

Galef, B. G. (1981). The ecology of weaning: Parasitism and the achievement of independence by altricial mammals. In *Parental Care in Mammals*, ed. D. J. Gubernick & P. H. Klopfer, pp. 211–41. New York: Plenum Press.

Galef, B. G. & Wigmore, S. W. (1983). Transfer of information concerning distant foods: A laboratory investigation of the 'information-centre' hypothesis. *Animal Behaviour*, **31**, 748–58.

Garner, R. L. (1892). *The Speech of Monkeys*. New York: Charles L. Webster.

Geist, V. (1966). The evolution of horn-like organs. *Behaviour*, **27**, 175–214.

Geist, V. (1974). On fighting strategies in animal combat. *Nature*, **250**, 354.

Gerhardt, H. C. (1974). The significance of some spectral features in mating call recognition in the green treefrog (*Hyla cinerea*). *Journal of Experimental Biology*, **61**, 229–41.

Gerhardt, H. C. (1987). Evolutionary and neurobiological implications of selective phonotaxis in the green treefrog, *Hyla cinerea*. *Animal Behaviour*, **35**, 1479–89.

Getty, T. (1996). Mate selection by repeated inspection: more on pied flycatchers. *Animal Behaviour*, **51**, 739–45.

Gibson, J. (1966). *The Senses Considered as Perceptual Systems*. Boston: Houghton Mifflin.

Gish, S. L. & Morton, E. S. (1981). Structural adaptations to local habitat acoustics in Carolina wren songs. *Zeitschrift für Tierpsychologie*, **56**, 74–84.

Godard, R. (1991). Long-term memory of individual neighbours in a migratory songbird. *Nature*, **350**, 228–9.

Gompertz, T. (1961). The vocabulary of the great tit. *British Birds*, **54**, 369–94.

Gompertz, T. (1967). The hiss-display of the great tit. *Vogelwelt*, **88**, 165–9.

Gottlander, K. (1987). Variation in the song rate of the male pied flycatcher (*Ficedula hypoleuca*): causes and consequences. *Animal Behavior*, **35**, 1037–43.

Gottlieb (1993). Social induction of malleability in ducklings: Sensory basis and psychological mechanism. *Animal Behaviour*, **45**, 707–19.

Gould, S. J. (1977). *Ontogeny and Phylogeny*. Cambridge: The Belknap Press of Harvard University Press.

Gouzoules, H. & Gouzoules, S. (1989). Design features and developmental modification of pigtail macaque, *Macaca nemestrina*, agonistic screams. *Animal Behaviour*, **37**, 383–401.

Gouzoules, S., Gouzoules, H. & Marler, P. (1984). Rhesus monkey (*Macaca mulatta*) screams: Representational signalling in the recruitment of agonistic aid. *Animal Behaviour*, **32**, 182–93.

Green, S. (1975). Variation of vocal pattern with social situation in the Japanese monkey (*Macaca fuscata*): A field study. In *Primate Behavior: Developments in Field and Laboratory Research*, ed. L. A. Rosenblum, pp. 1–102. New York: Academic Press.

Green, S. & Marler, P. (1979). The analysis of animal communication. In *Social Behavior and Communication. Handbook of Behavioral Neurobiology*, **3**, ed. P. Marler & J. G. Vandenbergh, pp. 73–158. New York: Plenum Press.

Greene, H. W. (1988). Antipredator mechanisms in reptiles. In *Biology of the Reptilia: Defense and Life History. Biology of the Reptilia*, **16**, ed. C. Gans & R. B. Huey, pp. 1–152. New York: Alan R. Liss.

Greenewalt, C. H. (1968). *Bird Song: Acoustics and Physiology*. Washington: Smithsonian.

Greenfield, M. D. (1994). Cooperation and conflict in the evolution of signal interactions. *Annual Review of Ecology and Systematics*, **25**, 97–126.

Greenwood, D. D. (1961). Auditory masking and the critical band. *Journal of the Acoustical Society of America*, **33**, 484–92.

Greig-Smith, P. W. (1980). Parental investment in nest defence by stonechats (*Saxicola torquata*). *Animal Behaviour*, **28**, 604–19.

Grinnell, J. & McComb, K. (1996). Maternal grouping as a defense against infanticide by males: evidence from field playback experiments on African lions. *Behavioral Ecology*, **7**, 55–9.

Guilford, T. & Dawkins, M. S. (1991). Receiver psychology and the evolution of animal signals. *Animal Behaviour*, **42**, 1–14.

Guilford, T. & Dawkins, M. S. (1992). Understanding signal design: a reply to Blumberg & Alberts. *Animal Behaviour*, **44**, 384–5.

Guilford, T. & Dawkins, M. S. (1993). Are warning colors handicaps? *Evolution*, **47**, 400–16.

Gustafson, G. E. & Green, J. A. (1989). On the importance of fundamental frequency and other acoustic features in cry perception and infant development. *Child Development*, **60**, 772–80.

Gustafson, G. E. & Green, J. A. (1991). Developmental coordination of cry sounds with visual regard and gestures. *Infant Behavior and Development*, **14**, 51–7.

Haggerty, T. M. & Morton, E. S. (1995). No. 188: Carolina wren (*Thryothorus ludovicianus*). In *The Birds of North America*, ed. A. Poole & F. Gill. Washington: The American Ornithologists' Union; Philadelphia: The Academy of Natural Sciences.

Hailman, J. P. (1977). *Optical Signals*. Bloomington: Indiana University Press.

Hailman, J. P., Haftorn, S. & Hailman, E. D. (1994). Male Siberian Tit *Parus cinctus* dawn serenades: suggestion for the origin of song. *Fauna Norvegica Series C, Cinclus*, **17**, 15–26.

Hamilton, W. D. (1964). The genetical evolution of social behaviour, I & II. *Journal of Theoretical Biology*, **7**, 1–52.

Hamilton, W. J. I. & McNutt, J. W. (1997). Determinants of conflict behavior. In *Communication. Perspectives in Ethology*, **12**, ed. D. H. Owings, M. D. Beecher & N. S. Thompson. New York: Plenum Press.

Hansen, A. J. (1986). Fighting behavior in bald eagles: A test of game theory. *Ecology*, **67**, 787–97.

Hansen, P. (1979). Vocal learning: its role in adapting sound structures to long-distance propagation, and a hypothesis on its evolution. *Animal Behaviour*, **27**, 1270–1.

Hanson, M. (1995). The Development of the California Ground Squirrels' Mammalian and Avian Antipredator Behavior. PhD, University of California, Davis.

Harper, D. G. C. (1991). Communication. In *Behavioural Ecology: An Evolutionary Approach*, ed. J. R. Krebs & N. B. Davies, pp. 374–97. Oxford: Blackwell Scientific Publications.

Harrington, F. H. (1987). Aggressive howling in wolves. *Animal Behaviour*, **35**, 7–12.

Harris, M. A. & Lemon, R. E. (1974). Songs of song sparrows: reactions of males to songs of different localities. *Condor*, **76**, 33–44.

Hartshorne, C. (1973). *Born to Sing*. Bloomington: Indiana University Press.

Hatfield, E., Cacioppo, J. T. & Rapson, R. L. (1994). *Emotional Contagion.* New York: Cambridge University Press.

Hauser, M. D. (1991). Sources of acoustic variation in rhesus macaque (*Macaca mulata*) vocalizations. *Ethology*, **89**, 29–46.

Hauser, M. D. (1996). *The Evolution of Communication.* Cambridge: MIT Press.

Hauser, M. D. (1997). Minding the behavior of deception. In *Machiavellian Intelligence II*, ed. A. Whiten & R. W. Byrne. Cambridge: Cambridge University Press.

Hauser, M. D. & Marler, P. (1993a). Food-associated calls in rhesus macaques (*Macaca mulatta*): I. Socioecological factors. *Behavioral Ecology*, **4**, 194–205.

Hauser, M. D. & Marler, P. (1993b). Food-associated calls in rhesus macaques (*Macaca mulatta*): II. Costs and benefits of call production and suppression. *Behavioral Ecology*, **4**, 206–12.

Hauser, M. D. & Nelson, D. A. (1991). 'Intentional' signaling in animal communication. *Trends in Ecology and Evolution*, **6**, 186–9.

Heinroth, O. (1911/1985). Contributions to the biology, especially the ethology and psychology of the Anatidae (English translation). In *Foundations of Comparative Ethology*, ed. G. M. Burghardt, pp. 246–301. New York: Van Nostrand Reinhold.

Hennessy, D. F. & Owings, D. H. (1988). Rattlesnakes create a context for localizing their search for potential prey. *Ethology*, **77**, 317–29.

Hennessy, D. F., Owings, D. H., Rowe, M. P., Coss, R. G. & Leger, D. W. (1981). The information afforded by a variable signal: Constraints on snake-elicited tail flagging by California ground squirrels. *Behaviour*, **78**, 188–226.

Herndon, J. G., Turner, J. J. & Collins, D. C. (1981). Ejaculation is important for mating-induced testosterone increases in male rhesus monkeys. *Physiology and Behavior*, **27**, 873–7.

Hersek, M. J. (1990). Behavior of predator and prey in a highly coevolved system: Northern Pacific rattlesnakes and California ground squirrels. PhD, University of California, Davis.

Hersek, M. J. & Owings, D. H. (1993). Tail flagging by adult California ground squirrels: a tonic signal that serves different functions for males and females. *Animal Behaviour*, **46**, 129–38.

Hersek, M. J. & Owings, D. H. (1994). Tail flagging by young California ground squirrels, *Spermophilus beecheyi*: age-specific participation in a tonic communicative system. *Animal Behaviour*, **48**, 803–11.

Hinde, R. A. (1981). Animal signals: ethological and games-theory approaches are not incompatible. *Animal Behaviour*, **29**, 535–42.

Hinde, R. A. (1985). Was 'The expression of the emotions' a misleading phrase? *Animal Behaviour*, **33**, 985–92.

Hinton, L., Nichols, J. & Ohala, J. J. (1994). *Sound Symbolism.* Cambridge: Cambridge University Press.

Hodl, W. (1977). Call differences and calling site segregation in anuran species from Central Amazonian floating meadows. *Oecologia*, **28**, 351–63.

Hofer, M. A. (1987). Early social relationships: a psychobiologist's view. *Child Development*, **58**, 633–47.

Hofer, M. A., Brunelli, S. A. & Shair, H. N. (1993). Ultrasonic vocalization responses of rat pups to acute separation and contact comfort do not depend on maternal thermal cues. *Developmental Psychobiology*, **26**, 81–95.

Hofer, M. A., Brunelli, S. A. & Shair, H. N. (1994). Potentiation of isolation-induced vocalization by brief exposure of rat pups to maternal cues. *Developmental Psychobiology*, **27**, 503–17.

Hoffman, M. L. (1978). Toward a theory of empathic arousal and development. In *The Development of Affect*, ed. M. Lewis & L. A. Rosenblum, pp. 227–56. New York: Plenum Press.

Hogstedt, G. (1983). Adaptation unto death: function of fear screams. *American Naturalist*, **121**, 562–70.

Hoi-Leitner, M., Nechtelberger, H. & Hoi, H. (1995). Song rate as a signal for nest site quality in blackcaps (*Sylvia atricapilla*). *Behavioral Ecology and Sociobiology*, **37**, 399–405.

Holekamp, K. E. & Sherman, P. W. (1989). Why male ground squirrels disperse. *American Scientist*, **77**, 232–9.

Hollis, K. L. (1984). The biological function of Pavlovian conditioning: The best defense is a good offense. *Journal of Experimental Psychology: Animal Behavior Processes*, **10**, 413–25.

Hollis, K. L. (1990). The role of Pavlovian conditioning in territorial aggression and reproduction. In *Contemporary Issues in Comparative Psychology*, ed. D. A. Dewsbury, pp. 197–219. Sunderland: Sinauer Associates.

Hoogland, J. L. (1983). Nepotism and alarm calling in the black-tailed prairie dog (*Cynomys ludovicianus*). *Animal Behaviour*, **31**, 472–9.

Hopkins, C. D. (1983). Sensory mechanisms in animal communication. In *Animal Behaviour, Vol. 2, Communication*, ed. T. R. Halliday & P. J. B. Slater, pp. 114-55. San Francisco: W. B. Freeman.

Hopp, S. L. & Morton, E. S. (1997). Sound playback studies. In *Animal Acoustic Communication: Sound Analysis and Research Methods*, ed. S. L. Hopp, M. J. Owren & C. S. Evans, pp. ??–??.. Heidelberg: Springer-Verlag.

Hopson, J. A. (1975). The evolution of cranial display structures in hadrosaurian dinosaurs. *Paleobiology*, **1**, 21–43.

Hopson, J. A. (1977). Relative brain size and behavior in archosaurian reptiles. *Paleobiology*, **1**, 21–43.

Horn, A. G. (1997). Speech acts and animal signals. In *Communication. Perspectives in Ethology*, **12**, ed. D. H. Owings, M. D. Beecher & N. S. Thompson. New York: Plenum Press.

Horner, J. R. & Makela, R. (1979). Nest of juveniles provides evidence of family structure among dinosaurs. *Nature*, **28**, 296–8.

Hughes, M. K., Hughes, A. L. & Covalt-Dunning, D. (1982). Stimuli eliciting foodcalling in domestic chickens. *Applied Animal Ethology*, **8**, 543–50.

Humphrey, N. K. (1976). The social function of intellect. In *Growing Points in Ethology*, ed. P. P. G. Bateson & R. A. Hinde, pp. 303–18. Cambridge: Cambridge University Press.

Huxley, J. (1923). Courtship activities in the red-throated diver (*Colymbus stellatus* Pontopp.); together with a discussion of the evolution of courtship in birds. *Journal of the Linnean Society of London*, **53**, 253–92.

Huxley, J. S. (1938). The present standing of the theory of sexual selection. In *Evolution: Essays on Aspects of Evolutionary Biology Presented to Professor E.S. Goodrich on his Seventieth Birthday*, ed. G.R. deBeer, pp. 11–42. Oxford: Clarendon Press.

Immelmann, K. (1969). On the effect of early experience upon sexual object fixation in estrildine finches. *Zeitschrift für Tierpsychologie*, **26**, 677–91.

Insley, S. J. (1996). Studies of Mother–Offspring Vocal Recognition in the Northern Fur Seal. PhD, University of California, Davis.

Johnston, R. F., Niles, D. M. & Rohwer, S. A. (1972). Hermon Bumpus and natural selection on the house sparrow, *Passer domesticus. Evolution*, **26**, 20–31.

Jürgens , U. (1979). Vocalizations as an emotional indicator, a neuroethological study in the squirrel monkey. *Behaviour*, **69**, 88–117.

Kacelnik, A. & Krebs, J. R. (1982). The dawn chorus in the great tit (*Parus major*): proximate and ultimate causes. *Behaviour*, **89**, 287–309.

Kamil, A. C. (1994). A synthetic approach to the study of animal intelligence. In *Behavioral Mechanisms in Evolutionary Ecology*, ed. L. A. Real, pp. 11–45. Chicago: University of Chicago Press.

Kaneyuki, H., Yokoo, H., Tsuda, A., Yoshida, M., Mizuki, Y., Yamada, M. & Tanaka, M. (1991). Psychological stress increases dopamine turnover selectively in mesoprefrontal dopamine neurons of rats: reversal by diazepam. *Brain Research*, **557**, 154–61.

Karakashian, S. J., Gyger, M. & Marler, P. (1988). Audience effects on alarm calling in chickens (*Gallus gallus*). *Journal of Comparative Psychology*, **102**, 129–35.

Kardong, K. V. (1986). Predatory strike behavior of the rattlesnake, *Crotalus viridis oreganus. Journal of Comparative Psychology*, **100**, 304–314.

Keverne, E. B., Martensz, N. D. & Tuite, B. (1989). Beta-endorphin concentrations in cerebrospinal fluid of monkeys are influenced by grooming relationships. *Psychoneuroendocrinology*, **14**, 155–61.

King, J. A. (1955). Social behavior, social organization, and population dynamics in a black-tailed prairiedog town in the Black Hills of South Dakota. *Contributions from the Laboratory of Vertebrate Biology, University of Michigan*, **67**, 1–123.

Klauber, L. M. (1940). A statistical study of the rattlesnakes. VII. The rattle, Part 1. *Occasional Papers of the San Diego Society for Natural History*, **6**, 1–62.

Klinnert, M. D., Campos, J. J., Sorce, J. F., Emde, R. N. & Svejda, M. (1983). Emotions as behavior regulators: Social referencing in infancy. In *The Emotion. Emotion in Early Development*, **2**, ed. R. Plutchik & H. Kellerman, pp. 57–86. New York: Academic Press.

Klump, G. (1996). Bird communication in the noisy world. In *Ecology and Evolution of Acoustic Communication in Birds*, ed. D. E. Kroodsma & E. H. Miller, pp. 321–38. Ithaca: Cornell University Press.

Klump, G. M., Kretzschmar, E. & Curio, E. (1986). The hearing of an avian predator and its avian prey. *Behavioral Ecology and Sociobiology*, **18**, 317–23.

Kodric-Brown, A. & Brown, J. H. (1984). Truth in advertising: the kinds of traits favored by sexual selection. *American Naturalist*, **124**, 309–23.

Konishi, M. (1965). The role of auditory feedback in the control of vocalization in the white-crowned sparrow. *Zeitschrift für Tierpsychologie*, **22**, 770–83.

Konishi, M. (1969). Time resolution by single auditory neurones in birds. *Nature*, **222**, 566–7.

Kraemer, G. W. (1992). A psychobiological theory of attachment. *Behavioral and Brain Sciences*, **15**, 493–541.

Krebs, J. R. (1977). The significance of song repertoires: the beau geste hypothesis. *Animal Behaviour*, **25**, 475–8.

Krebs, J. R., Ashcroft, R. & Orsdol, K. V. (1981). Song matching in the great tit, *Parus major*. *Animal Behaviour*, **29**, 919–23.

Krebs, J. R. & Dawkins, R. (1984). Animal signals: Mind reading and manipulation. In *Behavioural Ecology: An Evolutionary Appraoch*, ed. J. R. Krebs & N. B. Davies, pp. 380–402. Sunderland: Sinauer Associates.

Kroodsma, D. E. (1976). Reproductive development in a female songbird: differential stimulation by quality of male song. *Science*, **192**, 574–5.

Kroodsma, D. E. (1984). Songs of the alder flycatcher (*Empidonax alnorum*) and willow flycatcher (*Empidonax traillii*) are innate. *Auk*, **101**, 13–24.

Kroodsma, D. E. (1985). Development and use of two song forms by the eastern phoebe. *Wilson Bulletin*, **97**, 21–9.

Kroodsma, D. E. (1989). Male eastern phoebes (*Sayornis phoebe*; *Tyrannidae, Passeriformes*) fail to imitate songs. *Journal of Comparative Psychology*, **103**, 227–32.

Kroodsma, D. E. & Baylis, J. R. (1982). Appendix: a world survey of evidence for vocal learning in birds. In *Acoustic Communication in Birds*, **2**, ed. D. E. Kroodsma & E. H. Miller, pp. 311–37. New York: Academic Press.

Kroodsma, D. E. & Canady, R. A. (1985). Differences in repertoire size, singing behavior, and associated neuroanatomy among Marsh Wren populations have a genetic basis. *Auk*, **102**, 439–46.

Kroodsma, D.E. & Konishi, M. (1991). A suboscine bird (*Sayornis phoebe*) develops normal song without auditory feedback. *Animal Behavior*, **42**, 477–87.

Kroodsma, D. E. & Verner, J. (1987). Use of song repertoires among marsh wren populations. *Auk*, **104**, 63–72.

Lande, R. (1981). Models of speciation by sexual selection on polygenic characters. *Proceeding of the National Academy of Sciences, USA*, **78**, 3721–5.

Lazarus, R. S. (1991). *Emotion and Adaptation*. New York: Oxford University Press.

Leger, D. W. (1993). Contextual sources of information and responses to animal communication signals. *Psychological Bulletin*, **113**, 295–304.

Leger, D. W. & Owings, D. H. (1978). Responses to alarm calls by California ground squirrels: Effects of call structure and maternal status. *Behavioral Ecology and Sociobiology*, **3**, 177–86.

Lein, M. R. (1978). Song variation in a population of chestnut-sided warblers, its nature and suggested significance. *Canadian Journal of Zoology*, **56**, 1266–83.

Lemon, R. E. (1967). The response of cardinals to songs of different dialects. *Animal Behaviour*, **15**, 538–45.

Loffredo, C. A. & Borgia, G. (1986). Sexual selection, mating systems, and the evolution of avian acoustical displays. *American Naturalist*, **128**, 773–94.

Lorenz, K. (1970). *Konrad Lorenz: Studies of Human and Animal Behaviour (English translation by R D Martin)*. Cambridge: Harvard University Press.

Loughry, W. J. & McDonough, C. M. (1988). Calling and vigilance in California ground squirrels: A test of the tonic communication hypothesis. *Animal Behaviour*, **36**, 1533–40.

Lynch, A. (1996). The population memetics of birdsong. In *Ecology and Evolution of Acoustic Communication in Birds*, ed. D. E. Kroodsma & E. H. Miller, pp. 181–97. Ithaca: Cornell University Press.

Macedonia, J. M. (1990). What is communicated in the antipredator calls of lemurs: Evidence from playback experiments with ringtailed and ruffed lemurs. *Ethology*, **86**, 177–90.

Macedonia, J. M. & Evans, C. S. (1993). Variation among mammalian alarm call systems and the problem of meaning in animal signals. *Ethology*, **93**, 177–97.

Mainwaring, W. I. P., Haining, S. A. & Harper, B. (1988). The functions of testosterone and its metabolites. In *Hormones and their Actions*. 1, ed. B. A. Cooke, R. J. B. King & H. J. van der Molen, pp. 169–96. Amsterdam: Elsevier Scientific Publications.

Margoliash, D. (1983). Acoustic parameters underlying the responses of song-specific neurons in the white-crowned sparrow. *Journal of Neuroscience*, **3**, 1039–57.

Margoliash, D. (1986). Preference for autogenous song by auditory neurons in a song system nucleus of the white-crowned sparrow. *Journal of Neuroscience*, **6**, 1643–61.

Margoliash, D. & Fortune, E. S. (1992). Temporal and harmonic combination-sensitive neurons in the zebra finch HVc. *Journal of Neuroscience*, **12**, 4309–26.

Margoliash, D., Fortune, E. S., Sutter, M. L., Yu, A. C., Wren-Hardin, B. D. & Dave, A. (1994). Distributed representation in song system of oscines: evolutionary implications and functional consequences. *Brain, Behavior and Evolution*, **44**, 247–64.

Margoliash, D. & Konishi, M. (1985). Auditory representation of autogenous song in the song system of white-crowned sparrows. *Proceedings of the National Academy of Sciences USA*, **82**, 5997–6000.

Markl, H. (1985). Manipulation, modulation, information, cognition: Some of the riddles of communication. In *Experimental Behavioral Ecology and Sociobiology*, ed. B. Holldobler & M. Lindauer, pp. 163–94. Sunderland: Sinauer Associates.

Marler, P. (1955). Characteristics of some animal calls. *Nature*, **176**, 6–8.

Marler, P. (1956). The voice of the chaffinch and its function as language. *Ibis*, **98**, 231–61.

Marler, P. (1959). Developments in the study of animal communication. In *Darwin's Biological Work, Some Aspects Reconsidered*, ed. P. R. Bell, pp. 150-206. New York: John Wiley & Sons.

Marler, P. (1960). Bird songs and mate selection. In *Animal Sounds and Communication*, ed. W. E. Lanyon & W. N. Tavolga, pp. 348–67. Washington: American Institute of Biological Science.

Marler, P. (1961). The logical analysis of animal communication. *Journal of Theoretical Biology*, **1**, 295–317.

Marler, P. (1970). A comparative approach to vocal learning: Song development in white-crowned sparrows. *Journal of Comparative and Physiological Psychology*, **71**, 1–25.

Marler, P. (1975). On the origin of speech from language. In *The Role of Speech in Language*, ed. J. F. Kavanagh & J. E. Cutting, pp. 11–37. Cambridge: The MIT Press.

Marler, P. (1984). Animal communication: Affect or cognition? In *Approaches to Emotion*, ed. K. R. Scherer & P. Ekman, pp. 345–65. Hillsdale: Lawrence Erlbaum Associates.

Marler, P., Dufty, A. & Pickert, R. (1986). Vocal communication in the domestic chicken: I. Does a sender communicate information about the quality of a food referent to a receiver? *Animal Behaviour*, **34**, 188–93.

Marler, P., Evans, C. S. & Hauser, M. D. (1992). Animal signals: motivational, referential, or both? In *Nonverbal Vocal Communication: Comparative and Developmental Approaches*, ed. H. Papousek, U. Jurgens & M. Papousek, pp. 66-86. Cambridge: Cambridge University Press.

Marler, P. & Hamilton, W. J. (1966). *Mechanisms of Animal Behavior*. New York: John Wiley.

Marler, P., Karakashian, S. & Gyger, M. (1991). Do animals have the option of withholding signals when communication is inappropriate? The audience effect. In *Cognitive Ethology: The Minds of Other Animals (Essays in Honor of Donald R. Griffin)*, ed. C. A. Ristau, pp. 187–208. Hillsdale, NJ: Lawrence Erlbaum Associates.

Marler, P. & Peters, S. (1989). Species differences in auditory responsiveness in early vocal learning. In *The Comparative Psychology of Audition: Perceiving Complex Sounds*, ed. R. J. Dooling & S. H. Hulse, pp. 243–73. Hillsdale, NJ: Lawrence Erlbaum Associates.

Marler, P. & Tamura, M. (1962). Song 'dialects' in three populations of white-crowned sparrows. *Condor*, **64**, 368–77.

Marler, P. & Tamura, M. (1964). Culturally transmitted patterns of vocal behavior in sparrows. *Science*, **146**, 1483–6.

Martel, F. L., Nevison, C. M., Simpson, M. J. A. & Keverne, E. B. (1995). Effects of opioid receptor blockade on the social behavior of rhesus monkeys living in large family groups. *Developmental Psychobiology*, **28**, 71–84.

Marten, K. & Marler, P. (1977). Sound transmission and its significance for animal communication. I. Temperate habitats. *Behavioral Ecology and Sociobiology*, **2**, 271–90.

Marten, K., Quine, D. & Marler, P. (1977). Sound transmission and its significance for animal vocalization. II. tropical forest habitats. *Behavioral Ecology and Sociobiology*, **2**, 291–302.

Martin, G. (1981). Avian vocalizations and the sound interference model of Roberts et al. *Animal Behaviour*, **29**, 632–3.

Mason, W. A. (1971). Motivational factors in psychosocial development. In *Nebraska Symposium on Motivation*, ed. W. J. Arnold & M. M. Page, pp. 35–67. Lincoln: University of Nebraska Press.

Mason, W. A. (1979a). Environmental models and mental modes: Representational processes in the great apes. In *The Great Apes: Perspective on Human Evolution*, ed. D. A. Hamburg & E. R. McCown, pp. 277–93. Menlo Park, California: Benjamin/Cummings Publishing Co.

Mason, W. A. (1979b). Ontogeny of social behavior. In *Handbook of Behavioral Neurobiology, Vol. 3*, ed. P. Marler & J. G. Vandenbergh, pp. 1–28. New York: Plenum Press.

Mason, W. A. (1979c). Wanting and knowing: A biological perspective on maternal deprivation. In *Origins of the Infant's Social Responsiveness*, ed. E. Thoman, pp. 225–49. Hillsdale, NJ: Lawrence Erlbaum Associates.

Mason, W. A. (1986). Behavior implies cognition. In *Integrating Scientific Disciplines*, ed. W. Bechtel, pp. 297–307. Dordrecht, The Netherlands: Martinus Nijhoff Publishers.

Mateo, J. M. (1996). The development of alarm-call response behaviour in free-living juvenile Belding's ground squirrels. *Animal Behaviour*, **52**, 489–505.

Maynard Smith, J. (1974). The theory of games and the evolution of animal conflicts. *Journal of Theoretical Biology*, **47**, 209–21.

Maynard Smith, J. (1979). Game theory and the evolution of behaviour. *Proceedings of the Royal Society of London B*, **205**, 475–88.

Maynard Smith, J. (1982). *Evolution and the Theory of Games*. Cambridge: Cambridge University Press.

Maynard Smith, J. (1994). Must reliable signals always be costly? *Animal Behavior*, **47**, 1115–20.

Maynard Smith, J. & Price, G. R. (1973). The logic of animal conflict. *Nature*, **246**, 15–8.

Mayr, E. (1961). Causes and effect in biology. *Science*, **134**, 1501–6.

Mayr, E. (1963). *Animal Species and Evolution*. Cambridge: Harvard University Press.

Mayr, E. (1982). *The Growth of Biological Thought*. Cambridge: Belknap Press of Harvard University.

Mayr, E. (1988). *Toward a New Philosophy of Biology: Observations of an Evolutionist*. Cambridge: Belknap Press of Harvard University Press.

McConnell, P. B. (1991). Lessons from animal trainers: The effect of acoustic structure on an animal's response. In *Perspectives in Ethology*. **9**, ed. P. P. G. Bateson & P. H. Klopfer, pp. 165–87. New York: Plenum Press.

McGregor, P. K. (1991). The singer and the song: on the receiving end of bird song. *Biological Reviews*, **66**, 57–81.

McGregor, P. K. (1992). *Playback and Studies of Animal Communication*. New York: Plenum Press.

McGregor, P. K. (1994). Sound cues to distance: the perception of range. In *Perception and Motor Control in Birds*, ed. M. N. O. Davies & P. R. Green, pp. 74–94. Berlin: Springer-Verlag.

McGregor, P. K. & Avery, M. I. (1986). The unsung songs of great tits (*Parus major*): learning neighbour's songs for discrimination. *Behavioral Ecology and Sociobiology*, **18**, 311–16.

McGregor, P. K. & Dabelsteen, T. (1996). Communication networks. In *Ecology and Evolution of Acoustic Communication in Birds*, ed. D. E. Kroodsma & E. H. Miller, pp. 409–25. Ithaca: Cornell University Press.

McGregor, P. K. & Falls, R. B. (1984). The response of western meadowlarks (*Sturnella neglecta*) to the playback of undegraded and degraded calls. *Canadian Journal of Zoology*, **62**, 2125–8.

McGregor, P. K. & Krebs, J. R. (1984). Sound degradation as a distance cue in great tit (*Parus major*) song. *Behavioral Ecology and Sociobiology*, **16**, 49–56.

McGregor, P. K. & Krebs, J. R. (1988). Song learning in adult great tits (*Parus major*): effects of neighbours. *Behaviour*, **101**, 139–58.

McGregor, P. K., Krebs, J. R. & Ratcliffe, l. M. (1983). The response of great tits (*Parus major*) to the playback of degraded and undegraded songs: the effect of familiarity with the stimulus song type. *Auk*, **100**, 898–906.

Meier, V., Rasa, O. A. E. & Scheich, H. (1983). Call system similarity in a ground-living social bird and a mammal in the bush habitat. *Behavioral Ecology and Sociobiology*, **12**, 5–9.

Melchior, H. R. (1971). Characteristics of Arctic ground squirrel alarm calls. *Oecologia*, **7**, 184–90.

Michaels, C. F. & Carello, C. (1981). *Direct Perception*. Englewood Cliffs: Prentice-Hall.

Michelsen, A. (1978). Sound reception in different environments. In *In Sensory Ecology, Review and Perspectives*, ed. M. A. Ali, pp. 345–73. New York: Plenum Press.

Miller, J. G. (1965). Living systems: Basic concepts. *Behavioral Science*, **10**, 193–411.

Milligan, M. M. & Verner, J. (1971). Inter-population song dialect discrimination in the white-crowned sparrow. *Condor*, **73**, 208–13.

Mineka, S. & Cook, M. (1988). Social learning and the acquisition of snake fear in monkeys. In *Social Learning: Biological and Psychological Perspectives*, ed. T. R. Zentall & B. G. Galef, pp. 51–74. Hillsdale, NJ: Lawrence Erlbaum Associates.

Mineka, S., Davidson, M., Cook, M. & Keir, R. (1984). Observational conditioning of snake fear in rhesus monkeys. *Journal of Abnormal Psychology*, **93**, 355–72.

Mirsky, E. N. (1976). Song divergence in hummingbird and junco populations on Guadalupe Island. *Condor*, **78**, 230–2.

Mitchell, R. W. (1986). A framework for discussing deception. In *Deception: Perspectives on Human and Nonhuman Deceit*, ed. R. W. Mitchell & N. S. Thompson, pp. 3–40. Albany: State University of New York Press.

Møller, A. P. (1990). Deceptive use of alarm calls by male swallows, *Hirundo rustica*: A new paternity guard. *Behavioral Ecology*, **1**, 1–6.

Morse, D. H. (1974). Niche breadth as a function of social dominance. *American Naturalist*, **108**, 818–30.

Morse, D. H. (1967). The contexts of song in black-throated green and blackburnian warblers. *Wilson Bulletin*, **79**, 64–74.

Morton, E. S. (1970). Ecological sources of selection on avian sounds. PhD, Yale University.

Morton, E. S. (1975). Ecological sources of selection on avian sounds. *American Naturalist*, **109**, 17–34.

Morton, E. S. (1977). On the occurrence and significance of motivation-structural rules in some bird and mammal sounds. *American Naturalist*, **111**, 855–69.

Morton, E. S. (1982). Grading, discreteness, redundancy, and motivation-structural rules. In *Acoustic Communication in Birds*, **1**, ed. D. E. Kroodsma & E. H. Miller, pp. 183–212. New York: Academic Press.

Morton, E. S. (1986). Predictions from the ranging hypothesis for the evolution of long distance signals in birds. *Behaviour*, **99**, 65–86.

Morton, E. S. (1988). 'Innate': Outdated and inadequate or linguistic convenience? *Brain & Behavioral Sciences*, **11**, 642–3.

Morton, E. S. (1994). Sound symbolism and its role in non-human vertebrate communication. In *Sound Symbolism and Human Speech*, ed. L. Hinton, J. Ohala & J. Nichols, pp. 348–65. Cambridge: Cambridge University Press.

Morton, E. S. (1996a). Why songbirds learn songs: an arms race over ranging? *Poultry & Avian Biology Reviews*, **7**, 65–71.

Morton, E. S. (1996b). A comparison of vocal behavior among tropical and temperate passerine birds. In *The Evolution and Ecology of Vocal Behavior in Birds*, ed. D. E. Kroodsma & E. H. Miller, pp. 258–68. Ithaca: Cornell University Press.

Morton, E. S. & Derrickson, K. C. (1996). Song ranging by the dusky antbird, *Cercomacra tyrannina*: ranging without song learning. *Behavioral Ecology and Sociobiology*, **39**, 195–201.

Morton, E. S., Forman, L. & Braun, M. (1990). Extrapair fertilizations and the evolution of colonial breeding in purple martins. *Auk*, **107**, 275–83.

Morton, E. S., Gish, S. L. & Van der Voort, M. (1986a). On the learning of degraded and undegraded songs in the Carolina wren. *Animal Behaviour*, **34**, 815–20.

Morton, E. S. & Gonzalez, H. T. (1982). The biology of *Torreornis inexpectata* I. A comparison of vocalizations in *T. i. inexpectata and T. i. sigmani*. *Wilson Bulletin*, **94**, 433–46.

Morton, E. S., Lynch, J., Young, K. & Mehlhop, P. (1986b). Do male hooded warblers exclude females from nonbreeding territories in tropical forests? *Auk*, **104**, 133–5.

Morton, E. S. & Page, J. (1992). *Animal Talk: Science and the Voices of Nature*. New York: Random House.

Morton, E. S. & Shalter, M. D. (1977). Vocal response to predators in pair-bonded Carolina wrens. *Condor*, **79**, 222–7.

Morton, E. S. & Young, K. (1986). A previously undescribed method of song matching in a species with a single song 'type', the Kentucky warbler (*Oporornis formosus*). *Ethology*, **72**, 334–42.

Mountjoy, D. J. & Lemon, R. E. (1996). Female choice for complex song in the European starling: A field experiment. *Behavioral Ecology and Sociobiology*, **38**, 65–71.

Moynihan, M. (1970). Control, suppression, decay, disappearance and replacement of displays. *Journal of Theoretical Biology*, **29**, 85–112.

Moynihan, M. (1973). The evolution of behavior and the role of behavior in evolution. *Breviora*, **415**, 1–29.

Moynihan, M. (1982). Why is lying about intentions rare during some kinds of contests? *Journal of Theoretical Biology*, **97**, 7–12.

Moynihan, M. (1998). *The Social Regulation of Competition and Aggression: With a Discussion of Tactics and Strategies*. Washington: Smithsonian Institution Press.

Mundinger, P. C. (1982). Microgeographic and macrogeographic variation in acquired vocalizations of birds. In *Acoustic Communication in Birds*. **2**, ed. D. E. Kroodsma & E. H. Miller, pp. 147–200. New York: Academic Press.

Myrberg, A. A. (1981). Sound communication and hearing in fishes. In *Hearing and Sound Communication in Fishes*, ed. W. N. Tavolga, A. N. Popper & R. R. Fay, pp. 395–426. New York: Springer-Verlag.

Narins, P. M. & Capranica, R. R. (1976). Sexual differences in the auditory system of the treefrog Eleutherodactylus coqui. *Science*, **192**, 378–80.

Narins, P. M. & Smith, S. L. (1986). Clinal variation in anuran advertisement calls: basis for acoustic isolation? *Behavioral Ecology and Sociobiology*, **19**, 135–41.

Neil, S. J. (1983). Contests for space in breeding *Cichlasoma meeki*: The use of increased apparent size displays. *Behaviour*, **87**, 283–97.

Nelson, D. A. (1984). Communication of intentions during agonistic contexts by the pigeon guillemot, *Cepphus columbia*. *Behaviour*, **88**, 145–89.

Nelson, D. A., Marler, P. & Palleroni, A. (1995). A comparative approach to vocal learning: intraspecific variation in the learning process. *Animal Behaviour*, **50**, 83–97.

Neudorf, D. L., Stutchbury, B. J. M. & Piper, W. L. (1997). Covert extra-territorial behavior of female hooded warblers. *Behavioral Ecology*, **8**, ??–??.

Neuweiler, G. (1990). Auditory adaptations for prey capture in echolocating bats. *Physiological Reviews*, **70**, 615–41.

Nottebohm, F. (1972). The origins of vocal learning. *American Naturalist*, **106**, 116–40.

Nottebohm, F. (1975). Continental patterns of song variability in *Zonotrichia capensis* : some possible correlates. *American Naturalist*, **109**, 605–24.

Nottebohm, F. (1991). Reassessing the mechanisms and origins of vocal learning in birds. *Trends in Neuroscience*, **14**, 206–11.

Nottebohm, F. (1996). A white canary on Mount Acropolis. *Journal of Comparative Physiology*, **179**, 149–56.

Oetting, S., Prove, E. & Bischof, H.-J. (1995). Sexual imprinting as a two-stage process: Mechanisms of information storage and stabilization. *Animal Behaviour*, **50**, 393–403.

Ohala, J. J. (1980). The acoustic origin of the smile. *Journal of the Acoustic Society of America*, **68**, 33.

Ohala, J. J. (1984). An ethological perspective on common cross-language utilization of F_o of voice. *Phonetica*, **41**, 1–16.

Owings, D. H. (1994). How monkeys feel about the world: A review of how monkeys see the world. *Language and Communication*, **14**, 15–30.

Owings, D. H. & Coss, R. G. (1991). Context and animal behavior I: Introduction and review of theoretical issues. *Ecological Psychology*, **3**, 1–9.

Owings, D. H. & Hennessy, D. F. (1984). The importance of variation in sciurid visual and vocal communication. In *The Biology of Ground-Dwelling Squirrels: Annual Cycles, Behavioral Ecology, and Sociality*, ed. J. A. Murie & G. R. Michener, pp. 169–200. Lincoln: University of Nebraska Press.

Owings, D. H., Hennessy, D. F., Leger, D. W. & Gladney, A. B. (1986). Different functions of 'alarm' calling for different time scales: A preliminary report on ground squirrels. *Behaviour*, **99**, 101–16.

Owings, D. H. & Leger, D. W. (1980). Chatter vocalizations of California ground squirrels: Predator- and social-role specificity. *Zeitschrift für Tierpsychologie*, **54**, 163–84.

Owings, D. H. & Loughry, W. J. (1985). Variation in snake-elicited jump-yipping by black-tailed prairie dogs: Ontogeny and snake specificity. *Zeitschrift für Tierpsychologie*, **70**, 177–200.

Owings, D. H. & Morton, E. S. (1997). The role of information in communication: an assessment/management approach. In *Communication. Perspectives in Ethology*, **12**, ed. D. H. Owings, M. D. Beecher & N. S. Thompson. New York: Plenum Press.

Owings, D. H. & Owings, S. C. (1979). Snake-directed behavior by black-tailed prairie dogs (*Cynomys ludovicianus*). *Zeitschrift für Tierpsychologie*, **49**, 35–54.

Owings, D. H. & Virginia, R. A. (1978). Alarm calls of California ground squirrels (*Spermophilus beecheyi*). *Zeitschrift für Tierpsychologie*, **46**, 58–78.

Owren, M. J. & Rendall, D. (1997). An affective-conditioning model of nonhuman primate vocal signaling. In *Communication. Perspectives in Ethology*, **12**, ed. D. H. Owings, M. D. Beecher & N. S. Thompson. New York: Plenum Press.

Oyama, S. (1985). *The Ontogeny of Information: Developmental Systems and Evolution*. Cambridge: Cambridge University Press.

Palleroni, A. & Marler, P. (in press). Devolopment of anti-predator behaviour in galliforms is experience-dependant. *Animal Behaviour*,

Parker, G. A. (1974). Assessment strategy and the evolution of fighting behaviour. *Journal of Theoretical Biology*, **47**, 223–43.

Parker, G. A. & Rubenstein, D. I. (1981). Role assessment, reserve strategy, and acquisition of information in asymmetric animal contests. *Animal Behaviour*, **29**, 221–40.

Paton, D. (1986). Communication by agonistic displays: II. Perceived information and the definition of agonistic displays. *Behaviour*, **99**, 157–75.

Payne, R. B. (1979). Song structure, behavior, and sequence of song types in a population of village indigobirds, *Vidua chalybeata*. *Animal Behavior*, **27**, 997–1013.

Payne, R. B. (1982). Ecological consequences of song matching: breeding success and intraspecific song mimicry in indigo buntings. *Ecology*, **63**, 401–11.

Payne, R. B. (1981). Population structure and social behavior: models for testing the ecological significance of song dialects. In *Natural Selection and Social Behavior*, ed. R. D. Alexander & D. W. Tinkle, pp. 108–21. New York: Chiron Press.

Payne, R. B. (1996). Song traditions in indigo buntings: Origin, improvisation, dispersal, and extinction in cultural evolution. In *Ecology and Evolution of Acoustic Communication in Birds*, ed. D. E. Kroodsma & E. H. Miller, pp. 198–221. Ithaca: Cornell University Press.

Peirce, C. S. (1958). *The Collected Papers of Charles Sanders Peirce, 1931–1935*. Cambridge University Press.

Pepper, S. C. (1942). *World Hypotheses: A Study in Evidence*. Berkeley: University of California Press.

Pepperberg, I. M. (1992). What studies on learning can teach us about playback design. In *Paperback and Studies of Animal Communication*, ed. P. K. McGregor, pp. 47–57. New York: Plenum Press.

Pereira, M. E. & Macedonia, J. M. (1991). Ringtailed lemur anti-predator calls denote predator class, not response urgency. *Animal Behaviour*, **41**, 543–4.

Peters, G. (1984). On the structure of friendly close range vocalizations in terrestrial carnivores (*Mammalia, Carnivora, Fissipedia*). *Zeitschrift für Saugetierkunde*, **49**, 157–82.

Peters, G. (1989). Acoustic communication by fissiped carnivores. In *Carnivore Behavior, Ecology, and Evolution*, ed. J. L. Gittleman, pp. 14–56. Ithaca: Cornell University Press.

Philips, M. & Austad, S. N. (1990). Animal communication and social evolution. In *Interpretation and Explanation in the Study of Animal Behavior. Vol. 1: Interpretation, Intentionality, and Communication*, ed. M. Bekoff & D. Jamieson, pp. 254–68. Boulder: Westview Press.

Piaget, J. (1971). *Biology and Knowledge*. Chicago: University of Chicago Press.

Pickens, A. L. (1928). Auditory protective mimicry of the chickadee. *Auk*, **45**, 302–4.

Pinker, S. (1994). *The Language Instinct*. New York: Harper Collins.

Ploog, D. W. (1992). The evolution of vocal communication. In *Nonverbal Vocal Communication: Comparative and Developmental Approaches*, ed. V. Papousek, U. Jurgens & M. Papousek, pp. 6–30. Cambridge and Paris: Cambridge University Press and Editions de la Science de l'Homme.

Poole, J. H. (1987). Rutting behavior in African elephants: The phenomenon of musth. *Behaviour*, **102**, 283–316.

Poole, J. H. (1989a). Announcing intent: The aggressive state of musth in African elephants. *Animal Behaviour*, **37**, 140–52.

Poole, J. H. (1989b). Mate guarding, reproductive success and female choice in African elephants. *Animal Behaviour*, **37**, 842–49.

Poole, J. H., Kasman, L. H., Ramsay, E. C. & Lasley, B. L. (1984). Musth and urinary testosterone concentrations in the African elephant (*Loxodonta africana*). *Journal of Reproduction and Fertility*, **70**, 255–60.

Poole, J. H. & Moss, C. J. (1981). Musth in the African elephant, *Loxodonta africana*. *Nature*, **292**, 830–1.

Popp, J. W. & Ficken, R. W. (1987). Effects of non-specific singing on the song of the ovenbird. *Bird Behaviour*, **7**, 22–6.

Poran, N. S. & Coss, R. G. (1990). Development of antisnake defenses in California ground squirrels (*Spermophilus beecheyi*): I. Behavioral and immunological relationships. *Behaviour*, **112**, 222–45.

Pridmore-Brown, D. C. & Ingard, U. (1955). Sound propagation into the shadow zone in a temperature-stratified atmosphere above a plane boundary. *Journal of the Acoustical Society of America*, **27**, 36–42.

Prins, H. H. T. (1989). Condition changes and choice of social environment in African buffalo bulls. *Behaviour*, **108**, 297–324.

Proctor, H. C. (1993). Sensory exploitation and the evolution of male mating behaviour: A cladistic test using water mites. *Animal Behaviour*, **44**, 745–52.

Provine, R. R. (1996a). Laughter. *American Scientist*, **84**, 38–45.

Provine, R. R. (1996b). Contagious yawning and laughter: Significance for sensory feature detection, motor pattern generation, imitation, and the evolution of social behavior. In *Social Learning in Animals: The Roots of Culture*, ed. C. M. Heyes & B. G. Galef, pp. 179–208. San Diego: Academic Press.

Radesater, T., Jakobsson, S., Andbjer, N., Bylin, A. & Nystrom, K. (1987). Song rate and pair formation in the willow warbler, *Phylloscopus trochilus*. *Animal Behavior*, **35**, 1645–51.

Rand, A. S. & Ryan, M. J. (1981). The adaptive significance of a complex vocal repertoire in a neotropical frog. *Zeitschrift für Tierpsychologie*, **57**, 209–14.

Rasa, O. A. E. (1986). Coordinated vigilance in dwarf mongoose family groups: the 'watchman's song' hypothesis and the costs of guarding. *Ethology*, **71**, 340–4.

Ratcliffe, L. M. & Grant, P. R. (1985). Species recognition in Darwin's finches (Geospiza. Gould). III. Male responses to playback of different song types, dialects, and heterospecific songs. *Animal Behaviour*, **33**, 290–307.

Reid, M. L. (1987). Costliness and reliability in the singing vigour of Ipswich sparrows. *Animal Behavior*, **35**, 1735–43.

Rendall, D. (1996). Social communication and vocal recognition in free-ranging rhesus monkeys (*Macaca mulatta*). PhD, University of California, Davis.

Rendall, D., Rodman, P. S. & Emond, R. E. (1996). Vocal recognition of individuals and kin in free-ranging rhesus monkeys. *Animal Behaviour*, **51**, 1007–15.

Rescorla, R. A. (1988). Pavlovian conditioning: It's not what you think it is. *American Psychologist*, **43**, 151–60.

Richards, D. G. (1981). Estimation of distance of singing conspecifics by the Carolina wren. *Auk*, **98**, 127–33.

Richards, D. G. & Wiley, R. H. (1980). Reverberations and amplitude fluctuations in the propagation of sound in a forest: implications for animal communiction. *American Naturalist*, **115**, 381–99.

Ridley, M. (1983). *The Explanation of Organic Diversity: The Comparative Method and Adaptations for Mating*. Oxford: Clarendon Press.

Riechert, S. E. (1978). Games spiders play: Behavioral variability in territorial disputes. *Behavioral Ecology and Sociobiology*, **3**, 135–62.

Riechert, S. E. E. (1982). *Interaction strategies: Communication vs. coercion*. Princeton, New Jersey: Princeton University Press.

Roberts, J. P., Hunter, M. L. & Kacelnik, A. (1981). The ground effect and acoustic communication. *Animal Behaviour*, **29**, 633–4.

Rosenblatt, J. S. (1992). Hormone–behavior relations in the regulation of parental behavior. In *Behavioral Endocrinology*, ed. J. B. Becker, S. M. Breedlove & D. Crews, pp. 219–59. Cambridge: MIT Press.

Rothstein, S. I. (1975). Evolutionary rates and host defenses against avian brood parasites. *American Naturalist*, **109**, 161–76.

Rowe, M. P., Coss, R. G. & Owings, D. H. (1986). Rattlesnake rattles and burrowing owl hisses: A case of acoustic Batesian mimicry. *Ethology*, **72**, 53–71.

Rowe, M. P. & Owings, D. H. (1978). The meaning of the sound of rattling by rattlesnakes to California ground squirrels. *Behaviour*, **66**, 252–67.

Rowe, M. P. & Owings, D. H. (1990). Probing, assessment, and management during interactions between ground squirrels and rattlesnakes. Part 1: Risks related to rattlesnake size and body temperature. *Ethology*, **86**, 237–49.

Rowe, M. P. & Owings, D. H. (1996). Probing, assessment, and management during interactions between ground squirrels (*Rodentia: Sciuridae*) and rattlesnakes (*Squamata: Viperidae*). 2: Cues afforded by rattlesnake rattling. *Ethology*, **102**, 856–74.

Rowell, T. E. (1962). Agonistic noises of the Rhesus monkey. *Symposia of the Zoological Society of London*, **8**, 91-6.

Ryan, M. J. (1980). Female mate choice in a neotropical frog. *Science*, **209**, 523–5.

Ryan, M. J. (1985). *The Túngara Frog: A Study in Sexual Selection and Communication*. Chicago: University of Chicago Press.

Ryan, M. J. (1994). Mechanisms underlying sexual selection. In *Behavioral Mechanisms in Evolutionary Ecology*, ed. L. A. Real, pp. 190–215. Chicago: University of Chicago Press.

Ryan, M. J. & Keddy, H. A. (1992). Directional patterns of female mate choice and the role of sensory biases. **139**, S4–S35.

Ryan, M. J. & Rand, A.S. (1993a). Phylogenetic patterns of behavioral mate recognition systems in the *Physalaemus pustulosus* species group (Anura: Leptodactylidae): The role of ancestral and derived characters and sensory exploitation. In *Evolutionary Patterns and Processes*, ed. D. Lees & D. Edwards, pp. 251–67. London: Academic Press.

Ryan, M. J. & Rand, A. S. (1990). The sensory basis of sexual selection for complex calls in the Tungara frog, *Physalaemus pustulosus* (sexual selection for sensory exploitation). *Evolution*, **44**, 305–14.

Ryan, M. J. & Rand, A. S. (1993b). Sexual selection and signal evolution: The ghost of biases past. *Proceedings of the Royal Society, London*, **B, 340**, 187–95.

Ryan, M. J. & Rand, A. S. (1993c). Species recognition and sexual selection as a unitary problem in animal communication. *Evolution*, **47**, 647–57.

Ryan, M. J., Tuttle, M. D. & Rand, A. S. (1982). Bat predation and sexual advertisement in a neotropical frog. *American Naturalist*, **119**, 136–9.

Scherer, K. R. (1992). Vocal affect expression as symptom, symbol, and appeal. In *Nonverbal Vocal Communication: Comparative and Developmental Approaches*, ed. H. Papousek, U. Jurgens & V. Papousek, pp. 43–60. Cambridge: Cambridge University Press.

Schleidt, W. M. (1973). Tonic communication: continual effects of discrete signs in animal communication systems. *Journal of Theoretical Biology*, **42**, 359–86.

Schwagmeyer, P. L. & Foltz, D. W. (1990). Factors affecting the outcome of sperm competition in thirteen-lined ground squirrels. *Animal Behaviour*, **39**, 156–62.

Schwartz, J. J. & Wells, K. D. (1983a). An experimental study of acoustic interference between two species of neotropical treefrogs. *Animal Behaviour*, **31**, 181–90.

Schwartz, J. J. & Wells, K. D. (1983b). The influence of background noise on the behavior of a neotropical treefrog, *Hyla ebraccata*. *Herpetologica*, **39**, 121–9.

Schwartz, J. J. & Wells, K. D. (1984a). Vocal behavior of the neotropical treefrog *Hyla phlebodes*. *Herpetologica*, **40**, 452–63.

Schwartz, J. J. & Wells, K. D. (1984b). Interspecific acoustic interactions of the neotropical treefrog *Hyla ebraccata*. *Behavioral Ecology and Sociobiology*, **14**, 211–24.

Schwartz, J. J. & Wells, K. D. (1985). Intra- and interspecific vocal behavior of the neotropical treefrog *Hyla microcephala*. *Copeia*, 27–38.

Searcy, W. A. (1992). Song repertoire and mate choice in birds. *American Zoologist*, **32**, 71–80.

Sebeok, T. E. (1967). Discussion of communication processes. In *Social Communication Among Primates*, ed. S. A. Altmann, pp. 363–78. Chicago: University of Chicago Press.

Seyfarth, R. M. & Cheney, D. L. (1994). The evolution of social cognition in primates. In *Behavioral Mechanisms in Evolutionary Ecology*, ed. L. A. Real, pp. 371–89. Chicago: University of Chicago Press.

Seyfarth, R. M., Cheney, D. L. & Marler, P. (1980). Vervet monkey alarm calls: Semantic communication in a free-ranging primate. *Animal Behaviour*, **28**, 1070–94.

Shalter, M. D. (1978). Location of passerine seet and mobbing calls by goshawks and pygmy owls. *Zeitschrift für Tierpsychologie*, **46**, 260–7.

Shalter, M. D. & Schleidt, W. M. (1977). The ability of barn owls to discriminate and localize avian alarm calls. *Ibis*, **119**, 22–7.

Shannon, C. E. & Weaver, W. (1949). *The Mathematical Theory of Communication*. Urbana: University of Illinois Press.

Sherman, P. W. (1977). Nepotism and the evolution of alarm calls. *Science*, **197**, 1246-53.

Sherman, P. W. (1988). Levels of analysis. *Animal Behavior*, **36**, 616–19.

Shriner, W. M. (1995). Yellow-bellied marmot and golden-mantled ground squirrel responses to conspecific and heterospecific alarm calls. PhD Dissertation, University of California, Davis.

Shy, E. & Morton, E. S. (1986a). The role of distance, familiarity, and time of day in Carolina wrens' responses to conspecific songs. *Behavioral Ecology and Sociobiology*, **19**, 393–400.

Shy, E. & Morton, E. S. (1986b). Adaptation of amplitude structure of songs to propagation in field habitat in song sparrows. *Ethology*, **72**, 177–84.

Sieber, O. J. (1984). Vocal communication in raccoons (*Procyon lotor*). *Behaviour*, **90**, 80–113.

Silk, J. B., Cheney, D. L. & Seyfarth, R. M. (1996). The form and function of post-conflict interactions between female baboons. *Animal Behaviour*, **52**, 259–68.

Simon, H. A. (1994). The bottleneck of attention: connecting thought with motivation. In *Integrative Views of Motivation, Cognition, and Emotion. Nebraska Symposium on Motivation*, **41**, ed. W. D. Spaulding, pp. 1–21. Lincoln: University of Nebraska Press.

Slater, P. J. B. (1989). Birdsong learning: causes and consequences. *Ethology, Ecology, and Evolution*, **1**, 19–46.

Slater, P. J. B., Eales, L. A. & Clayton, N. S. (1988). Song learning in zebra finches (*Taeniopygia guttata*): Progress and prospects. In *Advances in the Study of Behavior*, **18**, ed. J. S. Rosenblatt, C. Beer, M.-C. Busnel & P. J. B. Slater, pp. 1–34. San Diego: Academic Press.

Smith, C. B. (1989). Risk-taking behavior in foraging juvenile herring gulls. MSc, University of Maryland, College Park.

Smith, W. J. (1963). Vocal communication of information in birds. *American Naturalist*, **97**, 117–26.

Smith, W. J. (1965). Message, meaning, and context in ethology. *American Naturalist*, **99**, 405–9.

Smith, W. J. (1977). *The Behavior of Communicating, An Ethological Approach*. Cambridge: Harvard University Press.

Smith, W. J. (1986a). An 'informational' perspective on manipulation. In *Deception, Perspectives on Human and Nonhuman Deceit*, ed. R. W. Mitchell & N. S. Thompson, pp. 71–86. Albany: State University of New York Press.

Smith, W. J. (1986b) Signaling behavior: Contributions of different repertoires. In *Dolphin Cognition and Behavior: A Comparative Approach*, ed. R. J. Schusterman, J. A. Thomas & F. G. Wood, pp. 315–30. Hillsdale, NJ: Lawrence Erlbaum Associates.

Smith, W. J. (1991a). Singing is based on two markedly different kinds of signaling. *Journal of Theoretical Biology*, **152**, 241–53.

Smith, W. J. (1991b). Animal communication and the study of cognition. In *Cognitive Ethology: The Minds of Other Animals*, ed. C. A. Ristau, pp. 209–30. Hillsdale, NJ: Lawrence Erlbaum Associates.

Smith, W. J. (1996). Using interactive playback to study how songs and singing contribute to communication about behavior. In *Ecology and Evolution of Acoustic Communication in Birds*, ed. D. E. Kroodsma & E. H. Miller, pp. 377–97. Ithaca: Cornell University Press.

Smith, W. J. (1997). The behavior of communicating, after twenty years. In *Communication. Perspectives in Ethology*, **12**, ed. D. H. Owings, M. D. Beecher & N. S. Thompson, pp. 7–53. New York: Plenum Press.

Smith, W. J., Pawlukiewicz, J. & Smith, S. (1978). Kinds of activities correlated with singing patterns of the yellow-throated vireo. *Animal Behaviour*, **26**, 862–84.

Smith, W. J., Smith, S. L., DeVilla, J. G. & Oppenheimer, E. C. (1976). The jump-yip display of the black-tailed prairie dog, *Cynomys ludovicianus*. *Animal Behaviour*, **24**, 609–21.

Smith, W. J., Smith, S. L., Oppenheimer, E. C. & DeVilla, J. G. (1977). Vocalizations of the black-tailed prairie dog, *Cynomys ludovicianus*. *Animal Behaviour*, **25**, 152–64.

Smotherman, W. P., Bell, R. W., Hershberger, W. A. & Coover, G. D. (1978). Orientation to rat pup cues: effects of maternal experiential history. *Animal Behaviour*, **26**, 265–73.

Snow, D. W. (1968). The singing assemblies of little hermits. *Living Bird*, **7**, 47–55.

Sorjonen, J. (1983). Transmission of the two most characteristic phrases of the song of the thrush nightingale *Lucinia lucinia* in different environmental conditions. *Ornis Scandinavia*, **14**, 278–88.

Squire, L. R. (1994). Declarative and nondeclarative memory: multiple brain systems supporting learning and memory. In *Memory Systems*, ed. D. L. Schacter & E. Tulving, pp. 203–31. Cambridge: MIT Press.

Stamps, J. (1995). Motor learning and the value of familiar space. *American Naturalist*, **146**, 41–58.

Steger, R. & Caldwell, R. L. (1983). Intraspecific deception by bluffing: A defense strategy of newly molted stomatopods (*Arthropoda: Crustacea*). *Science*, **221**, 558–60.

Stephens, D. W. & Krebs, J. R. (1986). *Foraging Theory*. Princeton: Princeton University Press.

Stevenson, J. G. (1967). Reinforcing effects of chaffinch song. *Animal Behaviour*, **15**, 427–32.

Stewart, K. J. & Harcourt, A. H. (1994). Gorillas' vocalizations during rest periods: signals of impending departure? *Behaviour*, **130**, 29–40.

Stiles, F. G. (1971). Time, energy, and territoriality of the Anna hummingbird (*Calypte anna*). *Science*, **173**, 818–21.

Stokes, A. W. (1971). Parental and courtship feeding in red jungle fowl. *Auk*, **88**, 21–9.

Strain, J. G. & Mumme, R. L. (1988). Effects of food supplementation, song playback, and temperature on vocal territorial beahvior of Carolina wrens. *Auk*, **105**, 11–16.

Stutchbury, B. J., Rhymer, J. & Morton, E. S. (1994). Parentage and plumage in hooded warblers: support for female control of extrapair fertilizations. *Behavioral Ecology*, **5**, 384–92.

Suboski, M. D. & Bartashunas, C. (1984). Mechanisms for social transmission of pecking preferences to neonatal chicks. *Journal of Experimental Psychology: Animal Behavior Processes*, **10**, 182–94.

Sullivan, B. K. (1982). Significance of size, temperature, and call attributes to sexual selection in *Bufo woodhousei australis*. *Journal of Herpetology*, **16**, 103–6.

Sullivan, K. A. (1984). Information exploitation by downy woodpeckers in mixed-species flocks. *Behaviour*, **91**, 294–311.

Swaisgood, R. R. (1994). Assessment of rattlesnake dangerousness by California ground squirrels. PhD, University of California, Davis.

Tamura, N. (1995). Postcopulatory mate guarding by vocalizations in the Formosan squirrel. *Behavioral Ecology and Sociobiology*, **36**, 377–86.

Tanaka, Y. (1996). Sexual selection enhances population extinction in a changing environment. *Journal of Theoretical Biology*, **180**, 197–206.

Tanner, W. P. & Swets, J. A. (1954). A decision-making theory of visual detection. *Psychological Review*, **61**, 401–9.

Temeles, E. J. (1994). The role of neighbours in territorial systems: when are they 'dear enemies'? *Animal Behavavior*, **47**, 339–50.

ten Cate, C. (1986). Does behavior contingent stimulus movement enhance filial imprinting in Japanese quail? *Developmental Psychobiology*, **19**, 607–14.

ten Cate, C. (1991). Behaviour-contingent exposure to taped song and zebra finch song learning. *Animal Behaviour*, **42**, 857–9.

ten Cate, C. (1994). Perceptual mechanisms in imprinting and song learning. In *Causal Mechanisms of Behavioural Development*, ed. J. A. Hogan & J. J. Bolhuis, pp. 116–46. Cambridge: Cambridge University Press.

Thompson, N. S. (1986). Ethology and the birth of comparative teleonomy. In *Relevance of Models and Theories in Ethology*, ed. R. Campan & R. Dayan, pp. 13–23, Toulouse, France: Privat, International Ethological Conference.

Thompson, N. S. (1997). Communication and natural design. In *Communication. Perspectives in Ethology*, **12**, ed. D. H. Owings, M. D. Beecher & N. S. Thompson. New York: Plenum Press.

Thompson, N. S., Olson, C. & Dessureau, B. (1996). Babies cries – who's listening – who's being fooled. *Social Research*, **63**, 763–84.

Thorpe, W. H. (1954). The process of song learning in the chaffinch, as studied by means of the sound spectrograph. *Nature*, **173**, 465.

Timberlake, W. & Lucas, G. A. (1989). Behavior systems and learning: From misbehavior to general principles. In *Contemporary Learning Theories: Instrumental Conditioning Theory and the Impact of Biological Constraints on Learning*, ed. S. B. Klein & R. R. Mowrer, pp. 237–75. Hillsdale, NJ: Lawrence Erlbaum Associates.

Tinbergen, N. (1951). *The Study of Instinct*. Oxford: Clarendon Press.

Tinbergen, N. (1952). Derived activities: Their causation, biological significance, origin and emancipation during evolution. *Quarterly Review of Biology*, **27**, 1–32.

Tinbergen, N. (1953). *The Herring Gull's World; a Study of the Social Behaviour of Birds*. London: Collins.

Tinbergen, N. (1959). Comparative studies of the behaviour of gulls (*Laridae*): a progress report. *Behaviour*, **15**, 1–70.

Tinbergen, N. (1963). On aims and methods of ethology. *Zeitschrift für Tierpsychologie*, **20**, 410–33.

Tinbergen, N. (1965). *Animal Behavior*. New York: Time-Life Books.

Toates, F. M. (1980). *Animal Behaviour : A Systems Approach*. New York: Wiley.

Tooby, J. & Cosmides, L. (1990). The past explains the present: Emotional adaptations and the structure of ancestral environments. *Ethology and Sociobiology*, **11**, 375–424.

Towers, S. (1987). Mimicry as a communicative process: Historical views and contemporary implications. Unpublished manuscript.

Trivers, R. L. (1972). Parental investment and sexual selection. In *Sexual Selection and the Descent of Man*, ed. B. Campbell, pp. 136–79. Chicago: Aldine

Trivers, R. L. (1974). Parent-offspring conflict. *American Zoologist*, **14**, 249–64.

Turner, G. F. & Huntingford, F. A. (1986). A problem for game theory analysis: Assessment and intention in male mouthbrooder contests. *Animal Behaviour*, **34**, 961–70.

Tyack, P. L. (1997). New directions in research on cetacean sonar. In *Communication. Perspectives in Ethology*, **12**, ed. D. H. Owings, M. D. Beecher & N. S. Thompson, New York: Plenum Press.

von Uexkull, J. (1909/1985). Environment (Umwelt) and inner world of animals (English translation). In *Foundations of Comparative Ethology*, ed. G. M. Burghardt, pp. 222–45. New York: Van Nostrand Reinhold Co.

Van Kampen, H. S. & Bolhuis, J. J. (1993). Interaction between auditory and visual learning during filial imprinting. *Animal Behaviour*, **45**, 623–5.

Van Valen, L. (1973). Body size and numbers of plants and animals. *Evolution*, **27**, 27–35.

Vencl, F. (1977). A case of convergence in vocal signals between marmosets and birds. *American Naturalist*, **111**, 777–82.

Vieth, W., Curio, E. & Ernst, U. (1980). The adaptive significance of avian mobbing: III. Cultural transmission of enemy recognition in blackbirds: Cross-species tutoring and properties of learning. *Animal Behaviour*, **28**, 1217–29.

Vos, D. R. (1994). Sex recognition in zebra finch males results from early experience. *Behaviour*, **128**, 1–14.

Waas, J. R. (1991). Do little blue penguins signal their intentions during aggressive interactions with strangers? *Animal Behaviour*, **41**, 375–82.

Wagner, R. H. (1993). The pursuit of extra-pair copulations by female birds: a new hypothesis of colony formation. *Journal of Theoretical Biology*, **163**, 333–46.

Waring, G. H. (1970). Sound communications of black-tailed, white-tailed, and Gunnison's prairie dogs. *American Midland Naturalist*, **83**, 167–85.

Waser, P. M. & Brown, C. H. (1986). Habitat acoustics and primate communication. *American Journal of Primatology*, **10**, 135–54.

Waser, P. M. & Waser, M. S. (1977). Experimental studies of primate vocalization: specializations for long-distance propagation. *Zeitschrift für Tierpsychologie*, **43**, 239–63.

Wasserman, F. E. (1977). Intraspecific acoustical interference in the white-throated sparrow, (*Zonotrichia albicollis*). *Animal Behaviour*, **25**, 949–52.

Watterson, T. & Riccillo, S. C. (1983). Vocal suppression as a neonatal response to auditory stimuli. *Journal of Auditory Research*, **23**, 205–14.

Weary, D. W. & Kramer, D. L. (1995). Response of eastern chipmunks to conspecific alarm calls. *Animal Behaviour*, **49**, 81–93.

Weeden, J. S. & Falls, J. B. (1959). Differential responses of male ovenbirds to recorded songs of neighboring and more distant individuals. *Auk*, **76**, 343–51.

Weishampel, D. B. (1981). Acoustic analyses of potential vocalizations in lambeosaurine dinosaurs (*Reptilia:Ornithischia*). *Paleobiology*, **7**, 252–61.

West, M. J. & King, A. P. (1985). Learning by performing: An ecological theme for the study of vocal learning. In *Issues in the Ecological Study of Learning. Resources for Ecological Psychology*, ed. T. D. Johnston & A. T. Pietrewicz, pp. 245–72. Hillsdale, NJ: Lawrence Erlbaum Associates

West, M. J. & King, A. P. (1990). Mozart's starling. *American Scientist*, **78**, 106–14.

West, M. & King, A. (1996). Eco-gen-actics: a systems approach to the ontogeny of avian communication. In *The Evolution and Ecology of Vocal Behavior in Birds*, ed. D. K. Kroodsma & E. H. Miller, pp. 20–38. Ithaca: Cornell University Press.

West, M. J., King, A. P. & Freeberg, T. M. (1994). The nature and nurture of neo-phenotypes: A case history. In *Behavioral Mechanisms in Evolutionary Ecology*, ed. L. A. Real, pp. 238–57. Chicago: University of Chicago Press.

West, M. J., King, A. P. & Freeberg, T. M. (1997). Building a social agenda for the study of bird song. In *Social Influences on Vocal Development*, ed. C. T. Snowdon & M. Hausberger, pp. 41–56. Cambridge: Cambridge University Press.

West-Eberhard, M. J. (1979). Sexual selection, social competition, and evolution. *Proceedings of the American Philosophical Society*, **123**, 222–34.

Wickler, W. (1968). *Mimicry in Plants and Animals*. New York: McGraw-Hill.

Wiley, R. H. (1975). Multidimensional variation in an avian display: implications for social communication. *Science*, **190**, 482–3.

Wiley, R. H. (1983). The evolution of communication: Information and manipulation. In *Animal Behaviour, Vol. 2: Communication*, ed. T. R. Halliday & P. J. B. Slater, pp. 156–89. Oxford: Blackwell Scientific Publications.

Wiley, R. H. & Godard, R. (1996). Ranging of conspecific songs by Kentucky warblers and its implications for interactions of territorial males. *Behaviour*, **133**, 81–102.

Wiley, R. H. & Richards, D. G. (1978). Physical constraints on acoustic communication in the atmosphere: implications for the evolution of animal vocalizations. *Behavioral Ecology and Sociobiology*, **3**, 69–94.

Wiley, R. H. & Richards, D. G. (1982). Adaptations for acoustic communication in birds: sound transmission and signal detection. In *Acoustic Communication in Birds*, **1**, ed. D. E. Kroodsma & E. H. Miller, pp. 131–70. New York: Academic Press.

Williams, G. C. (1966). *Adaptation and Natural Selection*. Princeton: Princeton University Press.

Williams, H. & Nottebohm, F. (1985). Auditory responses in avian vocal motor neurons: a motor theory for song perception in birds. *Science*, **229**, 279–82.

Wilson, E. O. (1975). *Sociobiology, the New Synthesis*. Cambridge: Belknap.

Wingfield, J. C. & Hahn, T. (1994). Testosterone and territorial behavior in sedentary and migratory sparrows. *Animal Behaviour*, **47**, 77–89.

Wise, R. A. (1987). Sensorimotor modulation and the variable action pattern (VAP): Toward a noncircular definition of drive and motivation. *Psychobiology*, **15**, 7–20.

Wynne-Edwards, V. C. (1962). *Animal Dispersion in Relation to Social Behaviour*. New York: Hafner.

Young, J. Z. (1954). Memory, heredity, and information. In *Evolution as a Process*, ed. J. Huxley, A. C. Hardy & E. B. Ford, pp. 281–99. London: Allen & Unwin.

Zahavi, A. (1975). Mate selection – a selection for a handicap. *Journal of Theoretical Biology*, **53**, 205–14.

Zahavi, A. (1977). Reliability in communication systems and the evolution of altruism. In *Evolutionary Ecology*, ed. B. Stonehouse & C. Perrins, pp. 253–260. London: MacMillan Press.

Zahavi, A. (1982). The pattern of vocal signals and the information they convey. *Behaviour*, **80**, 1–8.

Zahavi, A. (1987). The theory of signal selection and some of its implications. In *International Symposium of Biological Evolution*, ed. V. P. Delfino, pp. 305–27. Bari: Adriatic Editrice.

Zahavi, A. (1991). On the definition of sexual selection, Fisher's model, and the evolution of waste and of signals in general. *Animal Behaviour*, **42**, 501–3.

Zajonc, R. B. (1980). Feeling and thinking: Preferences need no inferences. *American Psychologist*, **35**, 151–75.

Zann, R. (1990). Song and call learning in wild zebra finches in south-east Australia. *Animal Behaviour*, **40**, 811–28.

Zuk, M. (1994). Immunology and the evolution of behavior. In *Behavioral Mechanisms in Evolutionary Ecology*, ed. L. A. Real, pp. 354–68. Chicago: University of Chicago Press.

Index